The Science of Rare Earth Elements

This book examines rare earth elements (REEs), materials, and metals that are critical to modern life. These serve as crucial ingredients in the latest technologies including electronics, electric motors, magnets, batteries, generators, energy storage systems (supercapacitors/pseudocapacitors), specialty alloys, and other emerging applications. REEs are used in various sectors including health care, transportation, power generation, petroleum refining, and consumer electronics. *The Science of Rare Earth Elements: Concepts and Applications* defines these elements, their histories, properties, and current and potential future applications across a wide range of industries across the world. It also discusses the environmental benefits, such as components in electric vehicles, wind turbines, solar applications, and energy storage systems. Conversely, the book also examines the liabilities of mining these REEs.

The Science of Rare Earth Elements

Concepts and Applications

Frank R. Spellman

CRC Press is an imprint of the
Taylor & Francis Group, an **informa** business

First edition published 2023
by CRC Press
6000 Broken Sound Parkway NW, Suite 300, Boca Raton, FL 33487-2742

and by CRC Press
4 Park Square, Milton Park, Abingdon, Oxon, OX14 4RN
CRC Press is an imprint of Taylor & Francis Group, LLC

Library of Congress Cataloging-in-Publication Data
Names: Spellman, Frank R., author.
Title: The science of rare earth elements : concepts and applications / Frank R. Spellman.
Description: First edition. | Boca Raton : CRC Press, 2023. | Includes bibliographical references and index.
Identifiers: LCCN 2022034102 (print) | LCCN 2022034103 (ebook) | ISBN 9781032396668 (hardback) | ISBN 9781032396682 (paperback) | ISBN 9781003350811 (ebook)
Subjects: LCSH: Rare earths. | Rare earth metals.
Classification: LCC TA418.9.R37 S64 2023 (print) | LCC TA418.9.R37 (ebook) | DDC 661.041—dc23/eng/20221011
LC record available at https://lccn.loc.gov/2022034102
LC ebook record available at https://lccn.loc.gov/2022034103

ISBN: 978-1-032-39666-8 (hbk)
ISBN: 978-1-032-39668-2 (pbk)
ISBN: 978-1-003-35081-1 (ebk)

DOI: 10.1201/9781003350811

Typeset in Times
by codeMantra

Contents

Preface

The Science of Rare Earth Elements: *For Renewable Energy Applications* is the seventh volume in the acclaimed series that includes *The Science of Water, The Science of Air, The Science of Environmental Pollution, The Science of Renewable Energy,* and *The Science of Waste* (in production) all of which bring this highly successful series fully into the twenty-first century. *The Science of Rare Earth Elements: Concepts and Applications* continues the series mantra based on good science and not feel-good science. It also continues to be presented in the author's trademark conversational style.

This book is about REEs, materials, and metals. REEs are critical to our modern way of life, although few people know or understand this. The truth be told this lack of knowledge or understanding of REEs is surprising because they are critical ingredients in todays' mix of technologies including electronics, electric motors, magnets, batteries, generators, energy storage systems (supercapacitors/pseudocapacitor s), emerging applications, and specialty alloys. REEs are used in various sectors of the US economy including health care, transportation, power generation, petroleum refining, and consumer electronics. This book focuses on answering the following questions: What are REEs? What is the history of REEs and their use? For what are REEs used? What are the properties of REEs? What are the environmental liabilities of mining REEs? And also, what does the future hold for the usage of REEs? Moreover, this book also asks the same basic and pertinent questions related to the topic of discussion: Why should we care about REEs?

This last question and the answer provided in the text is or should be of particular concern for those who are advocates for the use of renewable energy sources. Rare earth materials are extensively used in wind turbine operations to produce electrical power, in some solar applications for electrical power, and in energy storage systems. REEs are also used in electric vehicles, thereby decreasing the need to use fossil fuels for operation.

Concern for the environment and for the impact of environmental pollution has brought about the trend (and the need) to shift from the use and reliance on hydrocarbons to energy-power sources that are pollution neutral or near pollution neutral and renewable. We are beginning to realize that we are responsible for much of the environmental degradation of the past and present—all of which is readily apparent today. Moreover, the impact of 200 years of industrialization and surging population growth has far exceeded the future supply of hydrocarbon power sources. So, the implementation of renewable energy sources is surging, and along with it there is a corresponding surge in the utilization of rare earth materials for use in energy production.

Why a text on the science of REEs? Simply put, studying REEs, materials, metals, products, and so forth without including the inherent science connection is analogous to attempting to reach an unknown, unfamiliar location without being able to read a map, paper, or digital device.

Many of us have come to realize that a price is paid (sometimes a high price) for what is called "the good life." Our consumption and use of the world's resources make all of us at least partially responsible for pushing the need to prevent the pollution of our environment due to our use of conventional energy sources such as oil and coal. Pollution and its ramifications are one of the inevitable products of the good life we all strive to attain, but obviously pollution is not something caused by any single individual, nor can one individual totally prevent or correct the situation. The common refrain we hear today is to reduce pollution and its harmful effects everyone must band together as an informed, knowledgeable group and pressure the elected decision makers to manage the problem now and in the future. At this moment in time, there is an on-going push to substitute fossil fuels with renewable energy sources—this is where the shift to renewable energy sources comes into play and where the need to use REEs in wind turbines, solar technology, and energy storage applications is vital.

Throughout this text, common-sense approaches and practical examples have been presented. Again, because this is a science text, I have adhered to scientific principles, models, and observations, but you need not be a scientist to understand the principles and concepts presented. What is needed is an open mind, a love for the challenge of wading through all the information, an ability to decipher problems, and the patience to answer the questions relevant to each topic presented. The text follows a pattern that is nontraditional; that is, the paradigm used here is based on real-world experience, not on theoretical gobbledygook. Real-life situations are woven throughout the fabric of this text and presented in straightforward, plain English to give the facts, knowledge, and information to enable understanding and needed to make informed decisions.

Environmental issues are attracting ever-increasing attention at all levels. The problems associated with these issues are compounded and made more difficult by the sheer number of factors involved in managing any phase of any problem. Because the issues affect so many areas of society, the dilemma makes us hunt for strategies that solve the problems for all, while maintaining a safe environment without excessive regulation and cost—Gordian knots that defy easy solutions.

The preceding statement goes to the heart of why this text is needed. Presently, only a limited number of individuals have sufficient background in the science of REEs and their concepts and applications in the world of industrial and practical functions, purposes, and uses to make informed decisions on 21st-century product production, usage, and associated environmental issues.

Finally, *The Science of Rare Earth Elements* is designed to reach a wide range of practitioner and student backgrounds and also to provide a basic handbook or reference for wind and solar energy technicians, and those in other industries such as electric vehicle production and maintenance, those involved with the production of high tech devices such as smartphones, digital cameras, computer hard drives, light-emitting diodes (LEDs), fluorescent lights, flat-screen televisions, computer monitors, electronic displays, military applications, and trade secret operations whereby knowledge is fundamental to producing the trade secret operation. The text focuses on harnessing REEs to produce electrical power for transmission that is critical to preserving what we call the good life.

As in the past with the other editions in this series, this book is presented in the author's characteristic conversational style where the goal is to communicate with the reader and the user—failure to communicate is never an option with this writer—never.

The bottom line: Critical to solving these real-world environmental problems is for all of us to remember that old saying, we should take nothing but pictures, leave nothing but footprints, kill nothing but time and sustain ourselves with the flow of clean, safe, renewable energy.

Frank R. Spellman
Norfolk, VA

Conversion Factors and SI Units

The units most commonly used by environmental engineering professionals are based on the complicated English System of Weights and Measures. However, bench work is usually based on the metric system or the International System of Units (SI) due to the convenient relationship between milliliters (mL), cubic centimeters (cm^3), and grams (g).

The SI is a modernized version of the metric system established by the International agreement. The metric system of measurement was developed during the French Revolution and was first promoted in the United States in 1866. In 1902, proposed congressional legislation requiring the U.S. Government to use the metric system exclusively was defeated by a single vote. Although we use both systems in this text, SI provides a logical and interconnected framework for all measurements in engineering, science, industry, and commerce. The metric system is much simpler to use than the existing English system since all its units of measurement are divisible by 10.

Before listing the various conversion factors commonly used in environmental engineering it is important to describe the prefixes commonly used in the SI system. These prefixes are based on the power 10. For example, a "kilo" means 1,000 grams, and a "centimeter" means 100th of 1 meter. The 20 SI prefixes used to form decimal multiples and submultiples of SI units are given in Table 1.1.

Note that the kilogram is the only SI unit with a prefix as part of its name and symbol. Because multiple prefixes may not be used, in the case of the kilogram, the prefix names of SI prefixes are used with the unit's name "gram" and the prefix symbols are used with the unit symbol "g." With this exception, any SI prefix may be used with any SI unit, including the degree Celsius and its symbol °C.

SI Prefixes

Factor	Name	Symbol
10^{24}	Yotta	Y
10^{21}	Zetta	Z
10^{18}	Exa	E
10^{15}	Peta	P
10^{12}	Tera	T
10^{9}	Giga	G
10^{6}	Mega	M
10^{3}	Kilo	k
10^{2}	Hecto	h
10^{1}	Deka	da

(Continued)

10^{-1}	Deci	d
10^{-2}	Centi	c
10^{-3}	Milli	m
10^{-6}	Micro	μ
10^{-9}	Nano	n
10^{-12}	Pico	p
10^{-15}	Femto	f
10^{-18}	Atto	a
10^{-21}	Zepto	z
10^{-24}	Yocto	y

Example 1

10^{-6} kg = 1 mg (one milligram), but not 10^{-6} kg = 1 μkg (one microkilogram)

Example 2

Consider the height of the Washington Monument. We may write h_w = 169,000 mm = 16,900 cm = 169 m = 0.169 km using the millimeter (SI prefix "milli," symbol "m"); centimeter (SI prefix "centi," symbol "c"); or kilometer (SI prefix "kilo," symbol "k").

Example 3

Problem: Find degrees in Celsius of water at 72°F.

Solution:

$$°C = (°F - 32) \times 5/9 = (72 - 32) \times 5/9 = 22.2$$

DID YOU KNOW?

The Fibonacci sequence is the following sequence of numbers:

1, 1, 2, 3, 5, 8, 13, 21, 34, 55, 89, 144, …

Or, alternatively,

0.1, 1, 2, 3, 5, 8, 13, 21, 34, 55, 89, 144, …

Two important points, the first obvious: each term from the third onward is *the sum of the previous two.* Another point to notice is that if you divide each number in the sequence by the next number, beginning with the first, an interesting thing appears to be happening:

1/1 = 1, 1/2 = 0.5, 2/3 = 0.66666 …, 3/5 = 0.6, 5/8 = 0.625, 8/13 = 0.61538 …, 13/21 = 0.61904 …, (the first of these ratios appear to be converging to a number just a bit larger than 0.6).

CONVERSION FACTORS

Conversion factors are given below in alphabetical order and in unit category listing order.

ALPHABETICAL LISTING OF CONVERSION FACTORS

Factors	Metric (SI) or English Conversions
1 atm (atmosphere)=	1.013 bars
	10.133 N/cm^2 (Newtons/square centimeter)
	33.90 ft of H_2O (feet of water)
	101.325 kPa (kilopascals)
	1,013.25 mg (milligrams)
	13.70 psia (pounds/square inch—absolute)
	760 torr
	760 mm Hg (millimeters of mercury)
1 bar=	0.987 atm (atmospheres)
	1×10^6 dyn/cm^2 (dynes/square centimeter)
	33.45 ft of H_2O (feet of water)
	1×10^5 Pa [N/m^2] (newtons/square meter)
	750.06 torr
	750.06 mm Hg (millimeters of mercury)
1 Bq (becquerel)=	1 radioactive disintegration/second
	2.7×10^{-11} Ci (curie)
	2.7×10^{-8} mCi (millicurie)
1 BTU (British Thermal Unit)=	252 cal (calories)
	1,055.06 J (joules)
	10.41 liter-atmosphere
	0.293 watt-hours
1 cal (calories)=	3.97×10^{-3} BTUs (British Thermal Units)
	4.18 J (joules)
	0.0413 liter-atmospheres
	1.163×10^{-3} watt-hours
1 cm (centimeters)=	0.0328 ft (feet)
	0.394 in. (inches)
	10,000 μ (micrometers)
	100,000,000 Å = 10^8 Å (Ångstroms)
1 cc (cubic centimeter)=	3.53×10^{-5} ft^3 (cubic feet)
	0.061 in.^{3} (cubic inches)
	2.64×10^{-4} gal (gallons)
	52.18 L (liters)
	52.18 mL (milliliters)

(Continued)

Factors	Metric (SI) or English Conversions
1 ft^3 (cubic foot)=	28.317 cc (cubic centimeters)
	1,728 $in.^3$ (cubic inches)
	0.0283 m^3 (cubic meters)
	7.48 gal (gallons)
	28.32 L (liters)
	29.92 qts (quarts)
1 $in.^3$	16.39 cc (cubic centimeters)
	16.39 mL (milliliters)
	5.79×10^{-4} ft^3 (cubic feet)
	1.64×10^{-5} m^3 (cubic meters)
	4.33×10^{-3} gal (gallons)
	0.0164 L (liters)
	0.55 fl oz (fluid ounces)
1 m^3 (cubic meters)=	1,000,000 cc = 10^6 cc (cubic centimeters)
	33.32 ft^3 (cubic feet)
	61,023 $in.^3$ (cubic inches)
	264.17 gal (gallons)
	1,000 L (liters)
1 yd^3 (cubic yard)=	201.97 gal (gallons)
	764.55 L (liters)
1 Ci (curie)=	3.7×10^{10} radioactive disintegrations/second
	3.7×10^{10} Bq (becquerel)
	1,000 mCi (millicurie)
1 day=	24 hrs (hours)
	1,440 min (minutes)
	86,400 sec (seconds)
	0.143 weeks
	2.738×10^{-3} yrs (years)
1°C (expressed as an interval)=	1.8°F = [9/5]°F (degrees Fahrenheit)
	1.8°R (degrees Rankine)
	1.0 K (degrees Kelvin)
°C (degree Celsius)=	[(5/9)(°F – 32°)]
1°F (expressed as an interval)=	0.556°C = [5/9]°C (degrees Celsius)
	1.0°R (degrees Rankine)
	0.556 K (degrees Kelvin)
°F (degree Fahrenheit)=	[(9/5)(°C) + 32°]
1 dyn=	1×10^{-5} N (newton)
1 eV (electron volt)=	1.602×10^{-12} ergs
	1.602×10^{-19} J (joules)
1 erg=	1 dyne-centimeters
	1×10^{-7} J (joules)
	2.78×10^{-11} watt-hours
1 fps (feet/second)=	1.097 kmph (kilometers/hour)
	0.305 mps (meters/second)
	0.01136 mph (miles/hour)

(*Continued*)

Factors	Metric (SI) or English Conversions
1 ft (foot)=	30.48 cm (centimeters)
	12 in. (inches)
	0.3048 m (meters)
	1.65×10^{-4} nt. (nautical miles)
	1.89×10^{-4} mi. (statute miles)
1 gal (gallon)=	3,785 cc (cubic centimeters)
	0.134 ft^3 (cubic feet)
	231 $in.^3$ (cubic inches)
	3.785 L (liters)
1 g (gram)	0.001 kg (kilogram)
	1,000 mg (milligrams)
	1,000,000 ng = 10^6 ng (nanograms)
	2.205×10^{-3} lbs (pounds)
1 g/cc (grams/cubic centimeters)=	62.43 lbs/ft^3 (pounds/cubic foot)
	0.0361 lbs/$in.^3$ (pounds/cubic inch)
	8.345 lbs/gal (pounds/gallon)
1 Gy (gray)=	1 J/kg (joules/kilogram)
	100 rad
	1 Sv (sievert)—[unless modified through division by an appropriate factor, such as Q and/or N]
1 hp (horsepower)=	745.7 J/sec (joules/second)
1 hr (hour)=	0.0417 days
	60 min (minutes)
	3,600 sec (seconds)
	5.95×10^{-3} weeks
	1.14×10^{-4} yrs (years)
1 in. (inch)=	2.54 cm (centimeters)
	1,000 mils
1 inch of water=	1.86 mm Hg (millimeters of mercury)
	249.09 Pa
	0.0361 psi (lbs/$in.^2$)
1 J (joule)=	9.48×10^{-4} BTUs (British Thermal Units)
	0.239 cal (calories)
	10,000,000 ergs = 1×10^7 ergs
	9.87×10^{-3} liter-atmospheres
	1.0 N m (newton-meters)
1 kcal (kilocalories)=	3.97 BTUs (British Thermal Units)
	1,000 cal (calories)
	4,186.8 J (joules)
1 kg (kilogram)=	1,000 g (grams)
	2.205 lbs (pounds)
1 km (kilometer)=	3,280 ft (feet)
	0.54 nt. (nautical miles)
	0.6214 mi. (statute miles)

(Continued)

Factors	Metric (SI) or English Conversions
1 kW (kilowatt)=	56.87 BTUs/min (British Thermal Units/minute)
	1.341 hp (horsepower)
	1,000 J/sec (joules/second)
1 kW hr (kilowatt-hour)=	3,412.14 BTUs (British Thermal Units)
	3.6×10^6 J (joules)
	859.8 kcal (kilocalories)
1 L (liter)=	1,000 cc (cubic centimeters)
	1 dm^3 (cubic decimeters)
	0.0353 ft^3 (cubic feet)
	61.02 $in.^3$ (cubic inches)
	0.264 gal (gallons)
	1,000 mL (milliliters)
	1.057 qts (quarts)
1 m (meter)=	1×10^{10} Å (Ångstroms)
	100 cm (centimeters)
	3.28 ft (feet)
	39.37 in. (inches)
	1×10^{-3} km (kilometers)
	1,000 mm (millimeters)
	1,000,000 $\mu = 1 \times 10^6$ μ (micrometers)
	1×10^9 nm (nanometers)
1 mps (meters/second)=	196.9 fpm (feet/minute)
	3.6 kmph (kilometers/hour)
	2.237 mph (miles/hour)
1 mph (mile/hour)=	88 fpm (feet/minute)
	1.61 kmph (kilometers/hour)
	0.447 mps (meters/second)
1 nt. (nautical mile)=	6,076.1 ft (feet)
	1.852 km (kilometers)
	1.15 mi. (statute miles)
	2,025.4 yds (yards)
1 mi. (statute mile)=	5,280 ft (feet)
	1.609 km (kilometers)
	1,609.3 m (meters)
	0.869 nt. (nautical miles)
	1,760 yds (yards)
1 miCi (millicurie)=	0.001 Ci (curie)
	3.7×10^{10} radioactive disintegrations/second
	3.7×10^{10} Bq (becquerel)
1 mm Hg (millimeters of mercury)=	1.316×10^{-3} atm (atmosphere)
	0.535 in. H_2O (inches of water)
	1.33 mb (millibars)
	133.32 Pa
	1 torr
	0.0193 psia (pounds/square inch—absolute)

(*Continued*)

Factors	Metric (SI) or English Conversions
1 min (minute)=	6.94×10^{-4} days
	0.0167 hrs (hours)
	60 sec (seconds)
	9.92×10^{-5} weeks
	1.90×10^{-6} yrs (years)
1 N (newton)=	1×10^{5} dyn
1 N m (newton-meter)=	1.00 J (joules)
	2.78×10^{-4} watt-hours
1 ppm (parts/million-volume)=	1.00 mL/m^3 (milliliters/cubic meters)
1 ppm [wt] (parts/million-weight)=	1.00 mg/kg (milligrams/kilograms)
1 Pa=	9.87×10^{-6} atm (atmospheres)
	4.015×10^{-3} in. H_2O (inches of water)
	0.01 mb (millibars)
	7.5×10^{-3} mm Hg (millimeters of mercury)
1 lb (pound)=	453.59 g (grams)
	16 oz (ounces)
1 lbs/ft^3 (pounds/cubic foot)=	16.02 g/L (grams/liters)
1 lbs/ft^3 (pounds/cubic inch)=	27.68 g/cc (grams/cubic centimeter)
	1,728 lbs/ft^3 (pounds/cubic feet)
1 psi (pounds/square inch)=	0.068 atm (atmospheres)
	27.67 in. H_2O (inches or water)
	68.85 mb (millibars)
	51.71 mm Hg (millimeters of mercury)
	6,894.76 Pa
1 qt (quart)=	946.4 cc (cubic centimeters)
	57.75 in.^{3} (cubic inches)
	0.946 L (liters)
1 rad=	100 ergs/g (ergs/gram)
	0.01 Gy (gray)
	1 rem [unless modified through division by an appropriate factor, such as Q and/or N]
1 rem	1 rad [unless modified through division by an appropriate factor, such as Q and/or N]
1 Sv (sievert)=	1 Gy (gray) [unless modified through division by an appropriate factor, such as Q and/or N]
1 cm^2 (square centimeter)=	1.076×10^{-3} ft^2 (square feet)
	0.155 in.^{2} (square inches)
	1×10^{-4} m^2 (square meters)
1 ft^2 (square foot)=	2.296×10^{5} acres
	9.296 cm^2 (square centimeters)
	144 in.^{2} (square inches)
	0.0929 m^2 (square meters)
1 m^2 (square meter)=	10.76 ft^2 (square feet)
	1,550 in.^{2} (square inches)

(Continued)

Factors	Metric (SI) or English Conversions
1 mi.[2] (square mile)=	640 acres
	2.79×10^7 ft^2 (square feet)
	2.59×10^6 m^2 (square meters)
1 torr=	1.33 mb (millibars)
1 watt=	3.41 BTI/hr (British Thermal Units/hour)
	1.341×10^{-3} hp (horsepower)
	52.18 J/sec (joules/second)
1 watt-hour=	3.412 BTUs (British Thermal Units)
	859.8 cal (calories)
	3,600 J (joules)
	35.53 liter-atmosphere
1 week=	7 days
	168 hrs (hours)
	10,080 min (minutes)
	6.048×10^5 sec (seconds)
	0.0192 yrs (years)
1 yr (year)=	365.25 days
	8,766 hrs (hours)
	5.26×10^5 min (minutes)
	3.16×10^7 sec (seconds)
	52.18 weeks

CONVERSION FACTORS BY UNIT CATEGORY

Units of Length

1 cm (centimeter)=	0.0328 ft (feet)
	0.394 in. (inches)
	10,000 μ (micrometers)
	100,000,000 Å = 10^8 Å (Ångstroms)
1 ft (foot)=	30.48 cm (centimeters)
	12 in. (inches)
	0.3048 m (meters)
	1.65×10^{-4} nt. (nautical miles)
	1.89×10^{-4} mi. (statute miles)
1 in. (inch)=	2.54 cm (centimeters)
	1,000 mils
1 km (kilometer)=	3,280.8 ft (feet)
	0.54 nt. (nautical miles)
	0.6214 mi. (statute miles)

(Continued)

1 m (meter)=	1×10^{10} Å (Ångstroms)
	100 cm (centimeters)
	3.28 ft (feet)
	39.37 in. (inches)
	1×10^{-3} km (kilometers)
	1,000 mm (millimeters)
	1,000,000 μ = 1×10^6 μ (micrometers)
	1×10^9 nm (nanometers)
1 nt. (nautical mile)=	6,076.1 ft (feet)
	1.852 km (kilometers)
	1.15 mi. (statute miles)
	2.025.4 yds (yards)
1 mi. (statute mile)=	5,280 ft (feet)
	1.609 km (kilometers)
	1.690.3 m (meters)
	0.869 nt. (nautical miles)
	1,760 yds (yards)

Units of Area

1 cm^2 (square centimeter)=	1.076×10^{-3} ft^2 (square feet)
	0.155 in.^{2} (square inches)
	1×10^{-4} m^2 (square meters)
1 ft^2 (square foot)=	2.296×10^{-5} acres
	929.03 cm^2 (square centimeters)
	144 in.^{2} (square inches)
	0.0929 m^2 (square meters)
1 m^2 (square meter)=	10.76 ft^2 (square feet)
	1,550 in.^{2} (square inches)
1 mi.^{2} (square mile)=	640 acres
	2.79×10^7 ft^2 (square feet)
	2.59×10^6 m^2 (square meters)

Units of Volume

1 cc (cubic centimeter)=	3.53×10^{-5} ft^3 (cubic feet)
	0.061 in.^{3} (cubic inches)
	2.64×10^{-4} gal (gallons)
	0.001 L (liters)
	1.00 mL (milliliters)
1 ft^3 (cubic foot)=	28,317 cc (cubic centimeters)
	1,728 in.^{3} (cubic inches)
	0.0283 m^3 (cubic meters)
	7.48 gal (gallons)
	28.32 L (liters)
	29.92 qts (quarts)

(Continued)

1 in.^{3} (cubic inch)=	16.39 cc (cubic centimeters)
	16.39 mL (milliliters)
	5.79×10^{-4} ft^3 (cubic feet)
	1.64×10^{-5} m^3 (cubic meters)
	4.33×10^{-3} gal (gallons)
	0.0164 L (liters)
	0.55 fl oz (fluid ounces)
1 m^3 (cubic meters)=	1,000,000 cc = 10^6 cc (cubic centimeters)
	35.31 ft^3 (cubic feet)
	61,023 in.^{3} (cubic inches)
	264.17 gal (gallons)
	1,000 L (liters)
1 yd^3 (cubic yards)=	201.97 gal (gallons)
	764.55 L (liters)
1 gal (gallon)=	3,785 cc (cubic centimeters)
	0.134 ft^3 (cubic feet)
	231 in.^{3} (cubic inches)
	3.785 L (liters)
1 L (liter)=	1,000 cc (cubic centimeters)
	1 dm^3 (cubic decimeters)
	0.0353 ft^3 (cubic feet)
	61.02 in.^{3} (cubic inches)
	0.264 gal (gallons)
	1,000 mL (milliliters)
	1.057 qts (quarts)
1 qt (quart)=	946.4 cc (cubic centimeters)
	57.75 in.^{3} (cubic inches)
	0.946 L (liters)

Units of Mass

1 g (grams)=	0.001 kg (kilograms)
	1,000 mg (milligrams)
	1,000,000 mg = 10^6 ng (nanograms)
	2.205×10^{-3} lbs (pounds)
1 kg (kilogram)=	1,000 g (grams)
	2.205 lbs (pounds)
1 lbs (pound)=	453.59 g (grams)
	16 oz (ounces)

Units of Time

1 day=	24 hrs (hours)
	1440 min (minutes)
	86,400 sec (seconds)
	0.143 weeks
	2.738×10^{-3} yrs (years)
1 hr (hours)=	0.0417 days
	60 min (minutes)
	3,600 sec (seconds)
	5.95×10^{-3} yrs (years)
1 hr (hour)=	0.0417 days
	60 min (minutes)
	3,600 sec (seconds)
	5.95×10^{-3} weeks
	1.14×10^{-4} yrs (years)
1 min (minutes)=	6.94×10^{-4} days
	0.0167 hrs (hours)
	60 sec (seconds)
	9.92×10^{-5} weeks
	1.90×10^{-6} yrs (years)
1 week=	7 days
	168 hrs (hours)
	10,080 min (minutes)
	6.048×10^{5} sec (seconds)
	0.0192 yrs (years)
1 yr (year)=	365.25 days
	8,766 hrs (hours)
	5.26×10^{5} min (minutes)
	3.16×10^{7} sec (seconds)
	52.18 weeks

Units of the Measure of Temperature

°C (degrees Celsius)=	[(5/9)(°F – 32°)]
1°C (expressed as an interval)=	1.8°F = [9/5]°F (degrees Fahrenheit)
	1.8°R (degrees Rankine)
	1.0 K (degrees Kelvin)
°F (degree Fahrenheit)=	[(9/5)(°C) + 32°]
1°F (expressed as an interval)=	0.556°C = [5/9]°C (degrees Celsius)
	1.0°R (degrees Rankine)
	0.556 K (degrees Kelvin)

Units of Force

1 dyn=	1×10^{-5} N (newtons)
1 N (newton)=	1×10^{5} dyn (dynes)

Units of Work or Energy

1 BTU (British Thermal Unit)=	252 cal (calories)
	1,055.06 J (joules)
	10.41 liter-atmospheres
	0.293 watt-hours
1 cal (calories)=	3.97×10^{-3} BTUs (British Thermal Units)
	4.18 J (joules)
	0.0413 liter-atmospheres
	1.163×10^{-3} watt-hours
1 eV (electron volt)=	1.602×10^{-12} ergs
	1.602×10^{-19} J (joules)
1 erg=	1 dyne-centimeter
	1×10^{-7} J (joules)
	2.78×10^{-11} watt-hours
1 J (joule)=	9.48×10^{-4} BTUs (British Thermal Units)
	0.239 cal (calories)
	10,000,000 ergs = 1×10^{7} ergs
	9.87×10^{-3} liter-atmospheres
	1.00 N m (newton-meters)
1 kcal (kilocalorie)=	3.97 BTUs (British Thermal Units)
	1,000 cal (calories)
	4,186.8 J (joules)
1 kW hr (kilowatt-hour)=	3,412.14 BTUs (British Thermal Units)
	3.6×10^{6} J (joules)
	859.8 kcal (kilocalories)
1 N m (newton-meter)=	1.00 J (joules)
	2.78×10^{-4} watt-hours
1 watt-hour=	3.412 BTUs (British Thermal Units)
	859.8 cal (calories)
	3,600 J (joules)
	35.53 liter-atmospheres

Units of Power

1 hp (horsepower)=	745.7 J/sec (joules/second)
1 kW (kilowatt)=	56.87 BTUs/min (British Thermal Units/minute)
	1.341 hp (horsepower)
	1,000 J/sec (joules/second)
1 watt=	3.41 BTUs/hr (British Thermal Units/hour)
	1.341×10^{-3} hp (horsepower)
	1.00 J/sec (joules/second)

Units of Pressure

1 atm (atmosphere)=	1.013 bars
	10.133 N/cm^2 (newtons/square centimeter)
	33.90 ft of H_2O (feet of water)
	101.325 kPa (kilopascals)
	14.70 psia (pounds/square inch—absolute)
	760 torr
	760 mm Hg (millimeters of mercury)
1 bar=	0.987 atm (atmospheres)
	1×10^6 dyn/cm^2 (dynes/square centimeter)
	33.45 ft of H_2O (feet of water)
	1×10^5 Pa [N/m^2] (newtons/square meter)
	750.06 torr
	750.06 mm Hg (millimeters of mercury)
1 inch of water=	1.86 mm Hg (millimeters of mercury)
	249.09 Pa (pascals)
	0.0361 psi (lbs/in.^{2})
1 mm Hg (millimeter of mercury)=	1.316×10^{-3} atm (atmospheres)
	0.535 in. H_2O (inches of water)
	1.33 mb (millibars)
	133.32 Pa (pascals)
	1 torr
	0.0193 psia (pounds/square inch—absolute)
1 Pa (pascal)=	9.87×10^{-6} atm (atmospheres)
	4.015×10^{-3} in. H_2O (inches of water)
	0.01 mb (millibars)
	7.5×10^{-3} mm Hg (millimeters of mercury)
1 psi (pounds/square inch)=	0.068 atm (atmospheres)
	27.67 in. H_2O (inches of water)
	68.85 mb (millibars)
	51.71 mm Hg (millimeters of mercury)
	6,894.76 Pa (pascals)
1 torr=	1.33 mb (millibars)

Units of Velocity or Speed

1 fps (feet/second)=	1.097 kmph (kilometers/hour)
	0.305 mps (meters/second)
	0.01136 mph (miles/hour)
1 mps (meters/second)=	196.9 fpm (feet/minute)
	3.6 kmph (kilometers/hour)
	2.237 mph (miles/hour)
1 mph (mile/hour)=	88 fpm (feet/minute)
	1.61 kmph (kilometers/hour)
	0.447 mps (meters/second)

Units of Density

1 g/cc (grams/cubic centimeter)=	62.43 lbs/ft^3 (pounds/cubic foot)
	0.0361 lbs/in.^{3} (pounds/cubic inch)
	8.345 lbs/gal (pounds/gallon)
1 lbs/ft^3 (pounds/cubic foot)=	16.02 g/L (grams/liter)
1 lbs/in.^{2} (pounds/cubic inch)=	27.68 g/cc (grams/cubic centimeter)
	1.728 lbs/ft^3 (pounds/cubic foot)

Units of Concentration

1 ppm (parts/million-volume)=	1.00 mL/m^3 (milliliters/cubic meter)
1 ppm (wt)=	1.00 mg/kg (milligrams/kilogram)

Radiation and Dose Related Units

1 Bq (becquerel)=	1 radioactive disintegration/second
	2.7 × 10^{-11} Ci (curie)
	2.7 × 10^{-8} mCi (millicurie)
1 Ci (curie)=	3.7 × 10^{10} radioactive disintegration/second
	3.7 × 10^{10} Bq (becquerel)
	1,000 mCi (millicurie)
1 Gy (gray)=	1 J/kg (joule/kilogram)
	100 rad
	1 Sv (sievert)—[unless modified through division by an appropriate factor, such as Q and/or N]
1 mCi (millicurie)=	0.001 Ci (curie)
	3.7 × 10^{10} radioactive disintegrations/second
	3.7 × 10^{10} Bq (becquerel)

(Continued)

1 rad=	100 ergs/g (ergs/gram)
	0.01 Gy (gray)
	1 rem—[unless modified through division by an appropriate factor, such as Q and/or N]
1 rem=	1 rad—[unless modified through division by an appropriate factor, such as Q and/or N]
1 Sv (sievert)=	1 Gy (gray)—[unless modified through division by an appropriate factor, such as Q and/or N]

DID YOU KNOW?

Units and dimensions are not the same concepts. Dimensions are concepts like time, mass, length, weight, etc. Units are specific cases of dimensions, like hour, gram, meter, lb, etc. You can *multiply* and *divide* quantities with different units: 4 ft × 8 lb = 32 ft-lb; but you can *add* and *subtract* terms only if they have the same units: 5 lb +8 kg = **NO WAY!!!**

GEOLOGIC TIME SCALE

Erathem Or Era	System, Subsystem or Period, Subperiod	Series or Epoch
Cenozoic 65 million years ago to present "Age of Recent Life"	**Quaternary** 1.8 million years ago to the present	**Holocene** 11,477 years ago (+/–85 years) to the present—Greek "holos" (entire) and "ceno" (new).
		Pleistocene 1.8 million to approx. 11,477 (+/–85 years) years ago—The Great Ice Age—Greek words "pleistos" (most) and "ceno" (new).
	Tertiary 65.5 to 1.8 million years ago	**Pliocene** 5.3 to 1.8 million years ago—Greek "pleion" (more) and "ceno" (new).
		Miocene 23.0 to 5.3 million years ago—Greek "meion" (less) and "ceno" (new).

(*Continued*)

Erathem Or Era	System, Subsystem or Period, Subperiod	Series or Epoch	
		Oligocene 33.9 to 23.0 million years ago—Greek "oligos" (little, few) and "ceno" (new).	
		Eocene 55.8 to 33.9 million years ago—Greek "eos" (dawn) and "ceno" (new).	
		Paleocene 65.5 to 58.8 million years ago—Greek "palaois" (old) and "ceno" (new).	
Mesozoic 251.0 to 65.5 million years ago—Greek means "middle life"	**Cretaceous** 145.5 to 65.5 million years ago "The Age of Dinosaurs"	Late or upper Early or lower	
	Jurassic 199.6 to 145.5 million years ago	Late or upper Middle Early or lower	
	Triassic 251.0 in. 199.6 million years ago	Late or upper Middle Early or lower	
Paleozoic 542.0 to 251.0 million years ago, "Age of Ancient Life"	**Permian** 299.0 to 251.0 million years ago	Lopingian Guadalupian Cisuralian	**Devonian**
	Pennsylvanian 318.1 to 299.0 million years ago "The Coal Age"	Late or upper Middle Early or lower	
	Mississippian 359.2 to 318.1 million years ago	Late or upper Middle Early or lower	
	416.0 to 359.2 million years ago	Late or upper Middle Early or lower	
	Silurian 443.7 to 416.0 million years ago	Pridoli Ludlow Wenlock Llandovery	
	Ordovician 488.3 to 443.7 million years ago	Late or upper Middle Early or lower	
	Cambrian 542.0 to 488.3 million years ago	Late or upper Middle Early or Lower	
Precambrian Approximately 4 billion years ago to 542.0 million years ago			

About the Author

Frank R. Spellman, PhD, is a retired full-time adjunct assistant professor of environmental health at Old Dominion University, Norfolk, Virginia, and the author of more than 155 books covering topics ranging from concentrated animal feeding operations (CAFOs) to all areas of environmental science and occupational health. Many of his texts are readily available online at Amazon.com and Barnes and Noble.com, and several have been adopted for classroom use at major universities throughout the United States, Canada, Europe, and Russia; two have been translated into Spanish for South American markets. Dr. Spellman has been cited in more than 950 publications. He serves as a professional expert witness for three law groups and as an incident/accident investigator for the U.S. Department of Justice and a northern Virginia law firm. In addition, he consults on homeland security vulnerability assessments for critical infrastructures including water/wastewater facilities nationwide and conducts pre-Occupational Safety and Health Administration (OSHA)/Environmental Protection Agency EPA audits throughout the country. Dr. Spellman receives frequent requests to co-author with well-recognized experts in several scientific fields; for example, he is a contributing author of the prestigious text *The Engineering Handbook,* 2nd ed. (CRC Press). Dr. Spellman lectures on wastewater treatment, water treatment, homeland security, and safety topics throughout the country and teaches water/wastewater operator short courses at Virginia Tech (Blacksburg, Virginia). In 2011–2012, he traced and documented the ancient water distribution system at Machu Pichu, Peru, and surveyed several drinking water resources in Amazonia-Coco, Ecuador. He continues to collect and analyze contaminated sediments in the major river systems in the world. Dr. Spellman studied and surveyed two separate potable water supplies in the Galapagos Islands; he also studied and researched Darwin's finches while in the Galapagos. He holds a BA in public administration, a BS in business management, an MBA, and an MS and PhD in environmental engineering.

NOTE TO READER

The material presented in this book is based on the lectures I presented to my undergraduate and graduate students at Old Dominion University (ODU) during the 1999–2011 timeframe and on subsequent talks I presented at various national conferences. Again, this material is presented in my characteristic conversational style avoiding all gobbledygook and other long-winded and worthless tripe. I have found that in teaching, the object should not be to saturate the student with information, but instead to present information for discussion and for thought—presenting science or any other topic for consummation by and via a student's open mind allows him or her to think things out for themselves and this is what teaching at the college level should be all about—there is a lot to be said about a student's open mind.

Part I

The Foundation

1 The Vitamins of Modern Industry

In 1996, a graduate student in one of my environmental engineering classes asked me:

"How do you plan the books that you write?"

My answer:

"I write the book to find out what it reads like."

With the passage of time, my answer to this question remains the same.

F. Spellman

THE STANDARD QUESTIONS

Frank R. Spellman here, and welcome to Chapter 1 of our journey toward understanding rare earth elements (REEs) and their critical niche, role, and part in our future. This is a journey toward an understanding of what REEs are all about and throughout our trek you'll be importing information that is futuristic, but also germane to the now and present, the present and now. Anyway, let me say that in a number of the environmental health, environmental science, and environmental engineering classes I taught to undergraduate and graduate university students eventually, as part of the overall subject matter, it became time for me to present my various lectures on REEs and their impact not only on modern society but also on their potential effect on the environment.

So, on that first day or night when I delivered my first REEs lecture (there were several follow-up lectures, on-going, on the same topic but expanded—some would say presented in a quirky, off-the-way fashion—whatever) I waited until the students calmed down and glued their total attention to me or to the front of the class where I stood, focused, on what I was going to say and the eventual follow up (there is always follow up).

Anyway, I began my REE lecture with a question. I asked the following and at first only wanted a show of hands: "Okay, how many of you have heard of rare earth elements?"

Just about all of the 100 plus students raised their hands.

> Okay, that is good … now, how many of you can tell me how many rare earth elements there are?

Wow, I was surprised when everyone in the class answered that there are 15–17 REEs, depending on who is counting (note that all of my students were well-groomed in various levels of college chemistry—one of the prerequisites for entry into my classes).

DOI: 10.1201/9781003350811-2

Next, I asked: "Ok, how many of you can name each of the 15–17 REEs?"

There was an extended moment of silence.

Was I surprised? No. Not at all.

Before I could ask the next question about REEs, one of the grad students who was standing alongside the wall with several other students in the standing room-only classroom (all of the desks were taken) raised her hand and stated:

> Professor, because of the various levels of chemistry courses that I have taken I am familiar with many if not most of the REEs and know them when I see their names on the chart (i.e., the periodic table) but I am not certain how to pronounce them … I mean they mostly have strange names … hard to pronounce.

Was I surprised by her statement? No. Not at all.

Why?

Because she is and was correct, the REEs, unless you collaborate with them on a routine basis, have names that are not that easy to pronounce correctly.

Anyway, my next question for the overflowing number of students in the class was: "Okay, without having to name all of the REEs, and without pronouncing their names with perfect accuracy … instead, do any of you in the class here know what the REEs are used for?"

Katy bar the door! Well, a moment later when I recognized that this thought of mine was wrongly placed and used … and instead my thought turned to "great minds run in the same channel …" I was amazed, surprised, and pleased when over 100 hands were raised and shouts went out that their iPhones, laptops, and my mom's flatscreen TV all contained REEs and all this was spouted before I recognized a hand to speak forward … wow … that is how I felt. The only feeling of disappointment I felt was when no one put forward an answer stating that they are used in wind turbines, electric autos … renewable energy applications and so forth. However, I also knew that by the time my lecture series was over with (and had been absorbed) that the members of that large class, and other classes, would know and recognize the concepts and applications of REEs. They would know, understand, and accept that REEs are the vitamins of modern industry and the associated technology using REEs was soon to morph into technological advancements on steroids!

So, another standard question I am often asked while discussing REEs is "are rare earth elements really rare?"

Well, again, I expect this standard question, so I am prepared to provide an answer. What I say is that no, REEs are not rare. They are considered rare because they have rarely been mined because finding concentrations that are economically minable is unusual. Moreover, in the 18th and 19th centuries, they were named "rare earth elements" because most were identified as "earths" that could not be changed further by heat, and in comparison to other "earths" they were relatively rare. To make this point clear to the students I inform them that, for example, the REE cerium is the most abundant REE and is more common in the Earth's crust than copper or lead. All of the REEs, except promethium, are more abundant on average in the Earth's crust than silver, gold, or platinum. Finally, I turn on the projector-viewer and show on the large screen in front of the classroom a chart from USGS (2014) estimating the crustal abundance of REEs. I excluded promethium from the list because it is rare and unstable.

Rare Earth Element – Abundance in Earth's Crust [units of measure, parts per million (ppm)]

Rare Earth Element	Atomic Number	Crustal Abundance
Scandium	21	22
Yttrium	39	33
Lanthanum	57	39
Cerium	58	66.5
Praseodymium	59	9.2
Neodymium	60	41.5
Samarium	62	7.05
Europium	63	2
Gadolinium	64	6.2
Terbium	65	1.2
Dysprosium	66	5.2
Holmium	67	1.3
Erbium	68	3.5
Thulium	69	0.52
Ytterbium	70	3.2
Lutetium	71	0.8

For Degree of Rarity Comparison

Element	Atomic Number	Crustal Abundance
Gold	79	0.004
Silver	47	0.075
Lead	82	14
Copper	29	60

Source: USGS (2014). The Rate-Earth Elements—Vital to Modern Technologies and Lifestyles. Washington, DC: US Dept of Interior.

Note: This comparison between REEs and familiar metals like copper, lead, and gold is important because the rare earths are just as abundant as these. Figure 1.1 shows a few of these familiar metals and points out that gold, silver, palladium, platinum are the metals that when compared to REEs are truly rare metals, the exception being copper; it is more abundant than most REEs and is ranked only below the availability of cerium which is more abundant than copper.

FIGURE 1.1 This figure shows samples of the metals gold, silver, palladium, platinum, and copper, and—all of which, except copper, are rarer than REEs. Photo by F. Spellman.

The bottom line: When you are a teacher of smart, open minded, broad-minded, inquiring minds who ask pertinent questions, does it get any better than this?

SETTING THE FOUNDATION

I have a habit, a routine, a regular modus operandi of beginning the first lecture in all of my classes with definitions and acronyms, abbreviations of the terms, expressions, and technological labels or names used in the subject matter I am about to deliver.

So, because I adhere to those famous words of Voltaire—"if you wish to converse with me, please define your terms"—I start off with the key terms, abbreviations, acronyms, and definitions relative to the subject matter of the course I am teaching or information I am presenting. Now traditionalists would point out that definitions and so forth belong at the end of the book—in the glossary. My question is: why?

Note: Because REEs used in industry more than likely have some application related to magnetism and their magnetic properties, it is important to know or to be cognizant of magnetic terms, abbreviations, acronyms, and definitions used in magnetics.

DEFINITIONS, ABBREVIATIONS, AND ACRONYMS[1]

FCC—fluid catalytic cracking
RE—rare earth
REM—rare earth metals
REE—rare earth element

REO—rare earth oxide
REY—rare earth elements and yttrium
LREE—light rare earth elements
HREE—heavy rare earth elements
MREE—middle rare earth elements
MRI—magnetic resonance imaging
ppb—part per billion
ppt—part per trillion
ppm—part per million
Tesla (T)—the strength of magnetic flux density.
Gauss—a unit of magnetic induction, equal to one ten-thousandth of a tesla.

$$\text{Ampere's Law} - \sum B_{\parallel} \Delta l = \mu_0 I \quad (1.1)$$

where

B = strength of the magnetic field
l = length elements
μ_0 = permeability
I = current

Datum—unless otherwise stated, vertical and horizontal coordinate information is referenced to the World Geodetic System of 1984 (WGS 84). Altitude, as used in this book, refers to the distance above the vertical datum.

Concentration Unit	Equals
Milligram per kilogram (mg/kg)	Part per million
Microgram per gram (μg/g)	Part per million
Microgram per kilogram (μg/kg)	Part per billion (10^9)

EQUIVALENCIES

Part per million (ppm): 1 ppm = 1,000 ppb = 1,000.000 ppt = 0.0001%

Part per billion (ppb): 0.001 ppm = 1 ppb = 1,000 ppt = 0.0000001%

Part per trillion (ppt): 0.000001 ppm =0.001 ppb = 1 ppt = 0.0000000001%

Air gap—is a low permeability gap in the flux of a magnetic circuit. Often air, but inclusive of other materials such as aluminum, paint, and so forth.

Alkaline igneous rock—a series of igneous rocks that formed from magmas and fluids so enriched in alkali elements that sodium- and potassium-bearing form constituents of the rock in much greater proportion than normal igneous rocks.

Anisotropy—a property varying in magnitude according to the direction of measurement—magnets fall into this category and their anisotropy is developed by two methods: shape and magnetocrystalline.

Area of the air gap (AS)—measured in sq cm in a plane normal to the central flux line of the air par is the cross-sectional area of the air gap perpendicular to the flux path and is the average cross-sectional area of the portion of the air gap within which the application occurs.

Area of the magnet (Am)—measured in sq cm is the cross-sectional area of the magnet perpendicular to the flux line at any point along its length. Am is commonly referred to as the neutral section of the magnet.

B Magnetic Induction—the process in which a magnetic field produces a magnetic field induced by a field strength, H.

Bd Remanent Induction—measured in gauss or tesla Bd is the magnetic induction at any point on the demagnetization curve.

Bd/Hd—the permanence coefficient or shear line—it is the slope of the operating, the ratio of the Bd.

BdHd Energy Product—the energy product where the product of B·H maximum in kilojoules per cubic meter is called the maximum energy point.

Bg—magnetic induction in an airgap, measured in gauss.

BH Maximum Energy Product—BH max is the maximum product of BhHd which can be obtained in the demagnetization curve and is often used to denote grade—the grade value is especially used for rare earth magnets.

Bi (J) intrinsic induction—is the contribution of the magnetic material to the total magnetic induction, B.

Bm Recoil induction—is the magnetic induction that remains in the magnetic material after magnetizing; measured in gauss.

Bo Magnetic induction—at the point of maximum energy product$(BH)_{max}$; measure in gauss.

Br Residual induction (flux density)—is the magnetic induction corresponding to zero magnetizing force in a magnetic material after saturation in a closed circuit (i.e., closed circuit condition exists when the external flux path of a permanent magnet is confined with high permeability material); measured in gauss.

Carbonatite—a rare, carbonate igneous rock formed by magmatic or metasomatic processes. Most carbonatites consist of 50% or more primary carbonate minerals, such as calcite, dolomite, and ankerite. They are genetically associated with, and typically occur near, alkaline igneous rocks.

Domain, magnetic—areas within a magnet whereby all have the same orientation.

Eddy currents—are localized electric currents induced in a conductor by a varying magnetic field.

Electromagnet—(artificial magnet), composed of soft-iron cores around which are wound coils of insulated wire. When an electric current flows through the coil, the core becomes magnetized. When the current ceases to flow, the core loses most of the magnetism.

Epithermal—mineral veins and ore deposits formed within the Earths' crust from warm water at shallow depths at relatively low temperatures (50–200°C), generally at some distance from the magmatic source.

Ferromagnetic material—materials that can be magnetized by an external magnetic field and remain magnetized when magnetizing source is removed.

Flux—the *magnetic flux*, ϕ, (phi) is similar to current in the Ohm's Law formula and comprises the total number of lines of force existing in the magnetic circuit. The **Maxwell** is the unit of flux – that is, 1 line of force is equal to 1 Maxwell. *Note*: The **Maxwell** is often referred to as simply a line of force, line of induction, or line.

Hypabyssal—an igneous intrusion that solidified at shallow depths before reaching the Earth's surface.

Hysteresis—when the current in a coil of wire reverses thousands of times per second, a considerable loss of energy can occur. This loss of energy is caused by *hysteresis*. Hysteresis means "a lagging behind"; that is, the magnetic flux in an iron core lags behind the increases or decreases of the magnetizing force.

The simplest method of illustrating the property of hysteresis is by graphical means such as the hysteresis loop shown in Figure 1.2.

The hysteresis loop (Figure 1.2) is a series of curves that show the characteristics of a magnetic material. Opposite directions of current result in the opposite directions of +H and −H for field intensity. Similarly, opposite polarities are shown for flux density as +B and −B. The current starts at the center 0 (zero) when the material is unmagnetized. Positive H values increase B to saturation at $+B_{max}$. Next H decreases to zero, but B drops to the value of B, because of hysteresis. The current that produced the original magnetization now is reversed so that H becomes negative. B drops to zero and continues to $-B_{max}$. As the −H values decrease, B is reduced to −B, when H is zero. Now with a positive swing of current, H becomes positive, producing saturation at $+B_{max}$ again. The hysteresis loop is now completed. The curve does not return to zero at the center because of hysteresis.

Intensity—the unit of *intensity* of magnetizing force per unit of length is designated as H and is sometimes expressed as Gilberts per centimeter of length. Expressed as an equation,

$$H = \frac{NI}{L} \tag{1.2}$$

where

H = magnetic field intensity, ampere-turns per meter (At/m)
NI = ampere-turns, At
L = length between poles of the coil, m

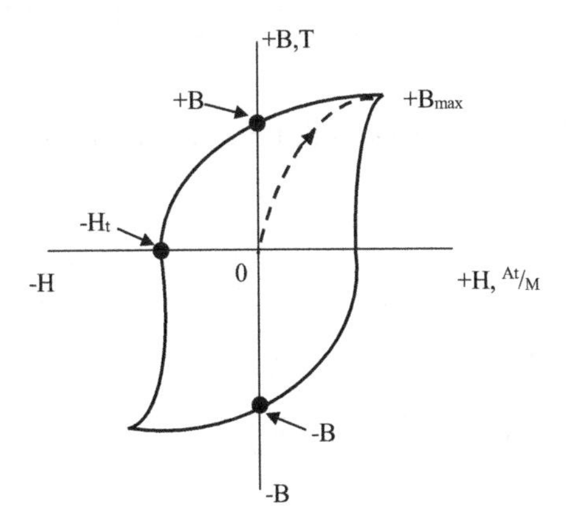

FIGURE 1.2 Hysteresis loop.

Oersted—In 1819, Hans Christian Oersted, a Danish scientist, discovered that a field of magnetic force exists around a single wire conductor carrying an electric current. In Figure 1.3, a wire is passed through a piece of cardboard and connected through a switch to a dry cell. With the switch open (no current flowing) if we sprinkle iron filings on the cardboard then tap it gently, the filings will fall back haphazardly. Now, if we close the switch, the current will begin to flow in the wire. If we tap the cardboard again, the magnetic effect of the current in the wire will cause the filings to fall back into a definite pattern of concentric circles with the wire as the center of the circles. Every section of the wire has this field of force around it in a plane perpendicular to the wire, as shown in Figure 1.4.

The ability of the magnetic field to attract bits of iron (as demonstrated in Figure 1.3) depends on the number of lines of force present. The strength of the magnetic field around a wire carrying a current depends on the current because it is the current that produces the field. The greater the current, the greater the strength of the field. A large current will produce many lines of force extending far from the wire, while a small current will produce only a few lines close to the wire, as shown in Figure 1.5.

Permeability—refers to the ability of a magnetic material to concentrate magnetic flux. Any material that is easily magnetized has high permeability. A measure of permeability for varied materials in comparison with air or vacuum is called

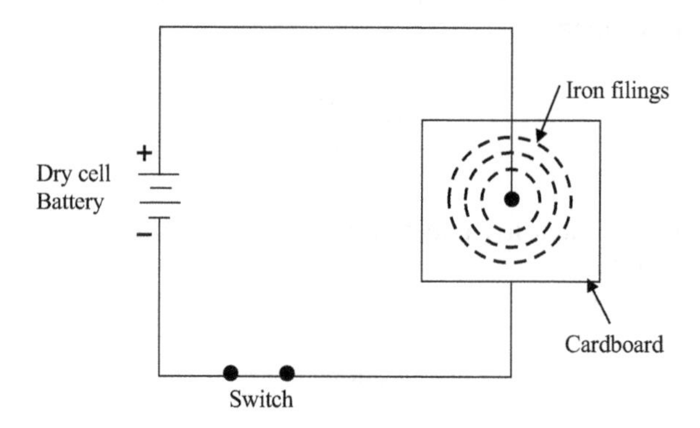

FIGURE 1.3 A circular patterns of magnetic force exists around a wire carrying an electric current.

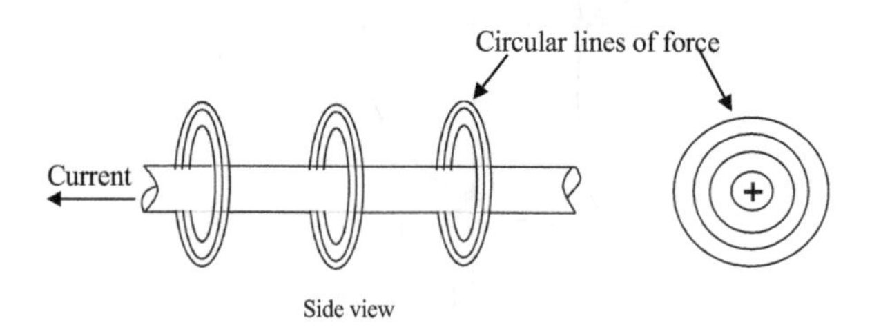

FIGURE 1.4 The circular fields of force around a wire carrying a current are in planes that are perpendicular to the wire.

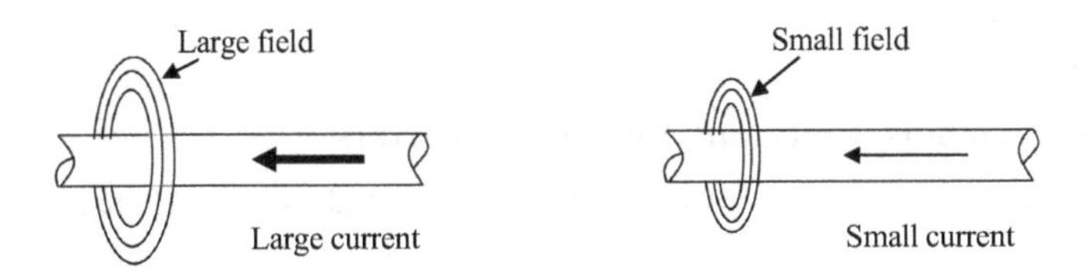

FIGURE 1.5 The strength of the magnetic field around a wire carrying a current depends on the amount of current.

relative permeability, symbolized by μ or (mu). When the core of an electromagnet is made of annealed sheet steel, it produces a stronger magnet than if a cast iron core is used. This is the case because the magnetizing force of the coil more readily acts upon annealed sheet steel than is the hard cast iron. Simply put, soft sheet steel is said to have greater *permeability* because of the greater ease with which magnetic lines are established in it. The permeability of air is arbitrarily set at 1. The permeability of other substances is the ratio of their ability to conduct magnetic lines compared to that of air. The permeability of nonmagnetic materials, such as aluminum, copper, wood, and brass, is unity, or the same as for air.

Reluctance—is analogous to resistance; it is the opposition to the production of flux in a material and is inversely proportional to permeability. Iron has high permeability and, therefore, low reluctance. Air has low permeability and hence high reluctance.

Sintered magnet—is comprised of a compacted powder which is then subjected to a heat treatment operation where the full density and magnetic orientation are achieved.

Strength of magnetic field—the *strength* of a magnetic field in a coil of wire depends on how much current flows in the turns of the coil. The more current, the stronger the magnetic field. In addition, the more turns, the more concentrated are the lines of force. The *force* that produces the flux in the magnetic circuit (comparable to electromotive force in Ohm's Law) is known as *Magnetomotive force*, or mmf. The practical unit of Magnetomotive force is the **ampere-turn** (At). In equation form,

$$F = \text{ampere-turns} = NI \tag{1.3}$$

where

F = Magnetomotive force, At

N = number of turns

I = current, A

Now that the major terminology used in magnetism has been listed and defined Table 1.1 lists the quantities symbols, units, and conversion factors used in magnetism.

S.I. = International System of Units

C.G.S. = Centimeter-gram-second system of units

Note: A quantity in S.I. units must be multiplied the ratios listed in the ration column to be converted to C.G.S. units.

Vickers hardness test (HV, Vickers Pyramid Number)—is a method to measure the hardness of materials (Smith and Sandland, 1922). The basic principle, as

TABLE 1.1
Quantities, Symbols, Units, and Conversion Factors

Quantity Column	Unit C.G.S.	Unit, S.I.	S.I./C.G.S. Ratio
Length l	Centimeter, cm	Meter, m	10^2
Mass	Gram, g	Kilogram, kg	10^3
Time t	Second, s	Second, s	I
Electric current I	Abampere	Ampere, A	10^{-1}
Temperature T	Degree Celsius, °C	Kelvin, K	K = °C + 273.16
Force	Dyne, dyn	Newton, N	10^5
Work or energy	Erg = dyn cm	Joule, J = Nm	10^7
Power	Erg/s	Watt, W = J/s	10^7
Magnet flux Φ	Maxwell	Weber, Wb	10^8
Flux density B	Gauss, G	Tesla, T = Wb/m²	10^4
Magnetic constant μ_0 (permeability of space)	Unity	Henry/meter, H m	$10^7/4\pi$
Intensity of magnetism J	e.m.u. = G/4π = dyn/cm² Oe	Tesla, T = N/Am	$10^4/4\pi$
Magnetic dipole moment j	e.m.u. = dyn cm/Oe	Wb m = Nm²/A	$10^{10}/4\pi$
Magnetic pole strength p	e.m.u.	Wb	$10^8/4\pi$
Magnetic field strength H	Oersted, Oe	Ampere/meter, Aim	$4\pi/10^3$
Magnetomotive force F	Gilbert, Gb	Ampere, A	$4\pi/10$
Permeability (abs.) $\mu = B/H$	–	Henry/meter, H/m	$10^7/4\pi$
Permeability (rel.) $\mu = B/\mu_0 H$			
Reluctance, R	Gilbert/maxwell	1/henry, H⁻¹	$4\pi/10^9$
Permeance (inverse of reluctance)	Maxwell/gilbert	Henry, H	$10^9/4\pi$
Susceptibility (rel.vol.)			
% = $J/\mu_0 H$	e.m.u.	Ratio	$1/4\pi$

with all common measures of hardness, is to observe a material's ability to resist plastic deformation from a standard source. The advantage of the Vickers test is that it can be used for all metals and has one of the broadest scales among hardness tests. The hardness number is determined by the load over the surface area of the indentation and not the area normal to the force, and it is thus not pressure.

FOUNDATION SET

My lectures on rare earth elements and other environmental topics typically last 3 hours, with a 15-minute break at the 1.5-hour point. All of the material presented to this point is presented during the first half of the class time. Although the material to this point is considered by many to be "dry and boring information" it is necessary foundational material before moving on to the material that follows.

So, it is time to replace the "dry" with the "wet" and the "boring" with the "interesting" because the foundation for further study is now set. Having said this, keep in mind that rare earth elements have been called the vitamins of modern industry and for good reason because they are used in so many modern applications especially as permanent magnets.

Rare-earth magnets are strong permanent magnets, the strongest type of permanent magnets, made from alloys of rare-earth elements. Although they have only been utilized since the 1970s, they have moved forward to a pinnacle place for practical usage in industry and elsewhere. The magnetic field typically produced by rare-earth magnets can exceed 1.4 teslas, whereas other common types of permanent magnets typically exhibit fields of 0.5 to 1 tesla.

So, with the foundational material presented, let's have a basic discussion about the properties of magnetism in general, and also of permanent magnets.

MAGNETISM

Most electrical equipment depends directly or indirectly upon magnetism. *Magnetism* is defined as a phenomenon associated with magnetic fields; that is, it has the power to attract such substances as iron, steel, nickel, or cobalt (metals that are known as magnetic materials). Correspondingly, a substance is said to be a magnet if it has the property of magnetism. For example, a piece of iron can be magnetized and is, thus, a magnet.

Can be magnetized—so, how does this occur?

To make a magnet magnetic it must be exposed to a strong external magnetic field or force. This field or force reorganizes the magnet's domain structure and leaves the magnet with a remanent magnetization (Br). If the magnet is isotropic (i.e., it has the same value when measured in different directions), the remanent magnetism has the same direction as the external field (the magnetic domains within the magnetized substance are aligned in the same direction—think of countless compass needles all pointing to a cardinal point on a compass). Meanwhile, an anisotropic magnet can only be magnetized in its anisotropy direction (i.e., it has different values when measured in different directions).

When magnetized, the piece of iron (*Note*: we will assume a piece of flat bar six inches long x 1-inch-wide x 0.5 inches thick; a bar magnet – see Figure 1.5) will have two points opposite each other, which most readily attract other pieces of iron. The points of maximum attraction (one on each end) are called the *magnetic poles* of the magnet: the north (N) pole and the south (S) pole. Just as like electric charges repel each other and opposite charges attract each other, like magnetic poles repel each other and unlike poles attract each other. Although invisible to the naked eye, its force can be shown to exist by sprinkling small iron filings on a glass covering a bar magnet as shown in Figure 1.6.

Figure 1.7 shows how the field looks without iron filings; it is shown as lines of force [known as *magnetic flux or flux lines*; the symbol for magnetic flux is the Greek lowercase letter ϕ (phi)] in the field, repelled away from the north pole of the magnet and attracted to its south pole.

Note: A *magnetic circuit* is a complete path through which magnetic lines of force may be established under the influence of a magnetizing force. Most magnetic circuits are composed of magnetic materials in order to contain the magnetic flux.

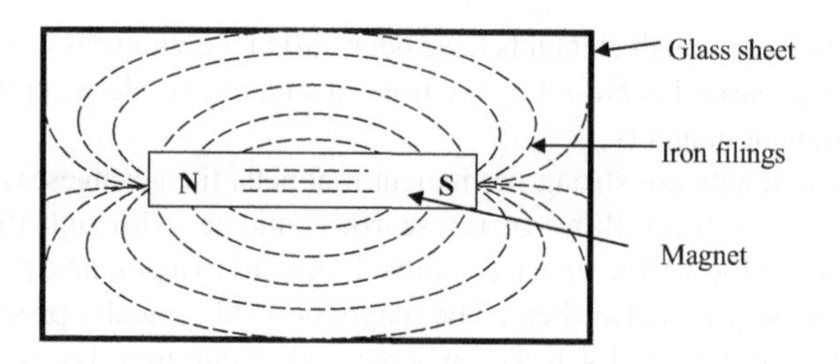

FIGURE 1.6 This figure shows the magnetic field around a bar magnet. If the glass sheet is tapped gently, the filings will move into a definite pattern that describes the field of force around the magnet.

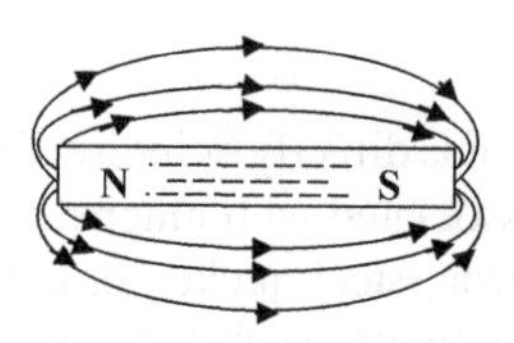

FIGURE 1.7 Magnetic field of force around a bar magnet, indicated by lines of force.

These circuits are similar to the electric circuit (a crucial point), which is a complete path through which current is caused to flow under the influence of an electromotive force.

There are three types or groups of magnets:

a. *Natural Magnets:* found in the natural state in the form of a mineral (an iron compound) called magnetite.
b. *Permanent Magnets:* (artificial magnet), hardened steel, or some alloy such as Alnico bars that have been permanently magnetized. The permanent magnet most people are familiar with is the horseshoe magnet (see Figure 1.8). Wind turbines use permanent magnets made from rare-earth materials; they are also used in several other applications.
c. *Electromagnets:* (artificial magnet), composed of soft-iron cores around which are wound coils of insulated wire. When an electric current flows through the coil, the core becomes magnetized. When the current ceases to flow, the core loses most of the magnetism.

Magnetic Materials

Natural magnets are no longer used (they have no practical value) in electrical circuitry because more powerful and more conveniently shaped permanent magnets can be produced artificially. Commercial magnets are made from special steels and alloys and rare-earth elements—magnetic materials.

Magnetic materials are those materials that are attracted or repelled by a magnet and that can be magnetized themselves. Iron, steel, and alloy bar are the most

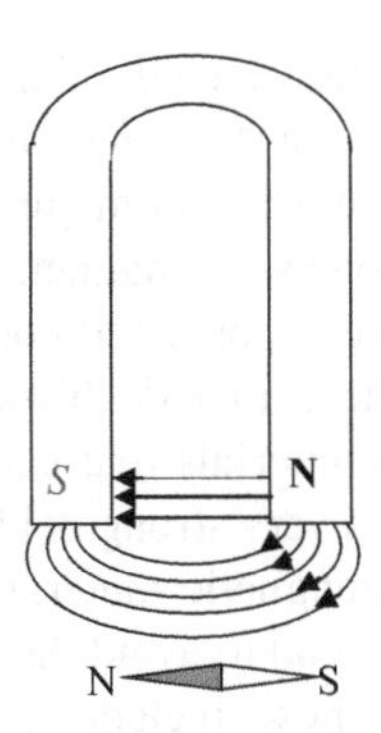

FIGURE 1.8 Permanent magnet.

common magnetic materials. These materials can be magnetized by inserting the material (in bar form) into a coil of insulated wired and passing a heavy direct current through the coil. The same material may also be magnetized if it is stroked with a bar magnet. It will then have the same magnetic property that the magnet used to induce the magnetism has—namely, there will be two poles of attraction, one at either end. This process produces a permanent magnet by *induction*—that is, the magnetism is induced in the bar by the influence of the stroking magnet.

Note: Permanent magnets are those of hard magnetic materials (durable steel or alloys) that retain their magnetism when the magnetizing field is removed. A temporary magnet is one that has no ability to retain a magnetized state when the magnetizing field is removed.

Even though classified as permanent magnets, it is important to point out that hardened steel and certain alloys are difficult to magnetize and are said to have a *low permeability* because the magnetic lines of force do not easily permeate or distribute themselves readily through the steel.

Note: *Permeability* refers to the ability of a magnetic material to concentrate magnetic flux. Any material that is easily magnetized has high permeability. A measure of permeability for varied materials in comparison with air or vacuum is called *relative* permeability, symbolized by μ or (mu).

Once durable steel and other alloys are magnetized, however, they retain a large part of their magnetic strength and are called *permanent magnets*. Conversely, materials that are easy to magnetize—such as soft iron and annealed silicon steel—are said to have a *high permeability*. Such materials retain only a small part of their magnetism after the magnetizing force is removed and are called *temporary magnets*.

The magnetism that remains in a temporary magnet after the magnetizing force is removed is called *residual magnetism*.

Early magnetic studies classified magnetic materials merely as being magnetic and nonmagnetic —that is, based on the strong magnetic properties of iron. However, because weak magnetic materials can be important in some applications, present studies classify materials into one of three groups: namely, paramagnetic, diamagnetic, and ferromagnetic.

a. *Paramagnetic materials*: These include aluminum, platinum, manganese, and chromium—materials that become only slightly magnetized even though under the influence of a strong magnetic field. This slight magnetization is in the same direction as the magnetizing field. Relative permeability is slightly more than 1 (i.e., considered nonmagnetic materials).
b. *Diamagnetic materials*: These include bismuth, antimony, copper, zinc, mercury, gold, and silver—materials that can also be slightly magnetized when under the influence of a very strong field. Relative permeability is less than 1 (i.e., considered nonmagnetic materials). When placed into a magnetic field, the lines of force tend to avoid the substance.
c. *Ferromagnetic materials*: These include iron, steel, nickel, cobalt, and commercial alloys—materials that are the most important group for applications of electricity and electronics. Ferromagnetic materials are easy to magnetize and have high permeability, ranging from 50 to 3,000. The lines of force tend to crowd into the material.

Magnetic Earth

The earth is a huge magnet; and surrounding earth is the magnetic field produced by the earth's magnetism. Most people would have no problem understanding or at least accepting this statement. However, if told that the earth's north magnetic pole is actually its south magnetic pole and that the south magnetic pole is actually the earth's north magnetic pole, they might not accept or understand this statement. However, in terms of a magnet, it is true.

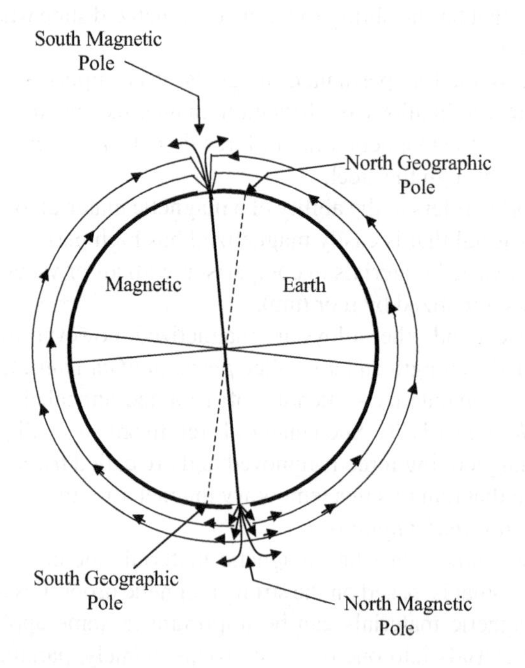

FIGURE 1.9 Earth's magnetic poles.

As can be seen from Figure 1.9, the magnetic polarities of the earth are indicated. The geographic poles are also shown at each end of the axis of rotation of the earth. As shown in Figure 1.9, the magnetic axis does not coincide with the geographic axis, and therefore the magnetic and geographic poles are not at the same place on the surface of the earth.

Recall that magnetic lines of force are assumed to emanate from the north pole of a magnet and to enter the South Pole as closed loops. Because the earth is a magnet, lines of force emanate from its north magnetic pole and enter the south magnetic pole as closed loops. A compass needle aligns itself in such a way that the earth's lines of force enter at its south pole and leave at its north pole. Because the north pole of the needle is defined as the end that points in a northerly direction, it follows that the magnetic pole near the north geographic pole is in reality a south magnetic pole, and vice versa.

So, are you confused, confounded, bewildered, baffled, or lost in some mystic magnetic field? Hang on, knowledge is enlightenment—and enlightenment is on the way.

NOTE

1 Based on material in F.R. Spellman (2000) *Basic Electricity*. Lancaster, PA: Technomic Publishing Co.; F.R. Spellman (2017) *The Science of Renewable Energy*. Boca Raton, FL: CRC Press; adaptation from Dura Magnetics (2021) *Glossary of Magnetic Terms*. Sylvania, OH.

REFERENCES

Smith, R.L. and Sandland, G.E. (1922). An accurate method of determining the harness of metals. *Proceedings of the Institution of Mechanical Engineers* 1: 623–641.

USGS (2014). Crustal Abundance of REE. USGS Mineral Resources Program. Accessed 12/12/21 @ https://pubs.gov/fs/2014/3078/pdf.

2 Basic Chemistry Review

INTRODUCTION

Chapter 1 presented a basic introduction to simple electrical circuits, electromagnetism, and a few other electrical properties and concepts, all of which are important to understanding the function, operation, and present purpose of utilizing rare earth elements in several different technical applications. So, at this point, it can be said that one of the fundamental building blocks allowing a clearer understanding of what rare-earth elements are all about is now in place. And now in this chapter, it is important to present another critical building block to enable understanding of the science involved with rare-metal mining, production, and usage. Although it is not the purpose of the brief, basic presentation in this chapter to make chemists out of anyone, it is essential for understanding the material that follows. Based on experience, I have found that when the basics, the nuts, and bolts of any branch of science are presented in ABC-fashion, understanding is right around the corner, so to speak. I remember one of my industrial hygiene students who when I asked that particular class what they thought of science or disciplines related to science, one young woman said that when she was younger and in high school she wanted no part of science, not because science is hard, but she wanted to learn something that she could understand, that she could grasp hold of without all the highfalutin, la-di-da, pretentious and snobbish bologna.

My reaction? Wow! I was totally taken aback by her comments about science in general. However, after a few minutes of digesting her comments, I realized that science is only difficult when it is presented in a fashion that is not clear to all. The key word is 'to all.' If you have attended classes or read a book whereby the content is presented in fast talk fashion or in short stubby sentences, and then carries on quickly to the next level of discussion, the question is how many listeners or readers got it? Another question: How many students initially understood the steppingstones to higher levels of understanding before being introduced to more complex topics?

The bottom line: When the student explained to me why she initially avoided science courses in school, I became the student and she the teacher. If we want to teach, to teach anything, we must be able to communicate to the listener or reader or both. By the way, when I had digested her comments I asked her, "So, you did not like science, but now you do? Why is that? Her answer: "I want a good paying job, occupation, position in life … so I bit the bullet so to speak and here I am."

ATOMS AND ELEMENTS

Just as we use several adjectives to describe an object (color of object, how tall or short, how bulky, or thin, how light, or heavy, etc.) several properties, or characteristics, must be used in combination to adequately describe a kind of matter. We

DOI: 10.1201/9781003350811-3

also distinguish one form of matter from another by its *properties*. However, simply saying that something is a colorless liquid isn't enough to identify it as alcohol. A lot of liquids are colorless, e.g., pure water, as well as many solutions. More details are needed before we can zero in on the identity of a substance. Chemists will, therefore, determine several properties, both chemical and physical, in order to characterize a particular sample of matter—to distinguish one form of matter from another. The following discussion describes the differences between the two kinds of properties, chemical and physical.

—F.R. Spellman (2021)

TOPICS COVERED

Defining Matter
Atomic Theory
Electron Configuration
Periodic Table of Elements

KEY TERMS USED IN THIS CHAPTER

Alkali metals—the elements in Group 1. They are very active metals that react with water to form strong bases.

Alkaline earth metals—the elements in Group 2 in the periodic table. Their oxides crumble and feel like dry earth.

Atom—the units of an element that make up molecules.

Atomic nucleus—the central part of an atom that contains the protons and neutrons

Atomic number—the number of protons in the nucleus

Atomic weight—the net weight of all protons and neutrons in the nucleus.

Electron—a particle that is part of all atoms. It has a negative electrical charge and almost no weight.

Element—substances whose atoms have the same atomic number.

Energy level—the orbit, or shell, in which the electron is located.

Gas—matter that does not occupy a definite volume and is shapeless.

Group—a family (vertical list) of elements in the periodic table.

Isotope—atoms with the same atomic number, but a different atomic weight.

Kinetic molecular theory—the idea that molecules are in constant motion.

Liquid—matter that occupies a definite volume but is shapeless.

Mass number—the sum of all protons and neutrons in the nucleus.

Melting point—the temperature at which a solid changes into a liquid.

Metal—an element whose atoms release valence electrons when they react.

Metalloids—elements that resemble both metals and nonmetals. They have properties found in metals and in nonmetals.

Molecule—the smallest particle of a substance that can exist with its own identity.

Neutrons—a particle that is a part of the atom. It has no electrical charge and has about the same weight as the proton.

Nonmetal—an element whose atoms gain valence electrons when they react.
Period—the horizontal listing of elements in the periodic table.
Periodic Table—the arrangement of all the elements by atomic numbers. Elements listed in vertical columns are in the same family because they share comparable properties.
Proton—a particle that is a part of all atoms. It has a positive electrical charge and weighs almost 2,000 times more than an electron.
Solid—matter that occupies a definite space (volume) and has a definite shape.

DEFINING MATTER

A thorough understanding of matter—how it consists of elements that are built from atoms—is critical for grasping chemistry. ***Matter*** is anything that occupies space and has weight (mass)—mass is responsible weight of the body, the larger the mass, the heavier the body. ***Matter*** (or mass–energy) is neither created nor destroyed during chemical change. So, what is weight? Weight is the measurement of earth's attraction to the body. Note that weight changes with the distance of the body from the center of the earth—the greater the distance the less the body weighs, and vice versa. Weight changes with altitude, but mass of the body is constant. For practical purposes, mass and weight are used interchangeably.

✓ **Key Point**: Matter is measured by making use of its two properties. Anything that has the properties of having weight and taking up space *must* be matter.

Along with the properties of having weight and taking up space, matter has two other distinct properties, *chemical* and *physical properties*. These properties are actually used to describe *substances*, which are definite varieties of matter. Copper, gold, salt, sugar, and rust are all examples of substances. All of these substances are uniform in their makeup. However, if we pick up a common rock from our garden, we cannot call the rock a substance because it is a mixture of several different substances.

✓ **Key Point**: A *substance* is a definite variety of matter, all specimens of which have the same properties.
✓ **Important Point**: It is interesting to note that under various environmental regulations, *chemical substances* are defined differently from the definition provided above. For example, in 40 CFR Section 710.2 chemical substance means any organic or inorganic substance of a particular molecular identity, including any combination of such substances occurring in whole or in part as a result of a chemical reaction or occurring in nature, and any chemical element or uncombined radical; except that chemical substance does not include: (1) Any mixture, (2) Any pesticide when manufactured, processed, or distributed in commerce for use as a pesticide, (3) Tobacco or any tobacco product, but not including any derivative products, (4) Any source material, special nuclear material, or byproduct material, (5) Any pistol, firearm, revolver, shells, and

cartridges, and (6) Any food, food additive, drug, cosmetic, or device, when manufactured, processed, or distributed in commerce for use as a food, food additive, drug, cosmetic, or device.

DID YOU KNOW?

An astronaut in space is weightless, although she has the same mass as on earth.

PHYSICAL PROPERTIES

Substances have two kinds of physical properties: intensive and extensive. *Intensive physical properties* include those features that definitely distinguish one substance from another. Intensive physical properties do not depend on the amount of the substance. Some of the important intensive physical properties are color, taste, melting point, boiling point, density, luster, and hardness. It is important to note that it takes a combination of several intensive properties to identify a given substance. For example, a certain substance may have a particular color that is common to it, but not necessarily unique to it. A white diamond is, as its name implies, white. However, is another gemstone that is white and faceted to look like a diamond really a diamond? Remember that a diamond is one of the hardest known substances. To determine if the white faceted gemstone is really a diamond, we would also have to test its hardness and not rely on its appearance alone.

Some of the important intensive physical properties are defined as follows:

1. **Density**—the mass per unit volume of a substance. Suppose we had a cube of lard and a large box of crackers, each having a mass of 400 g. The density of the crackers would be much less than the density of the lard because the crackers occupy a much larger volume than the lard occupies.

   The density of an object or substance can be calculated by using the formula:

$$\text{Density} = \text{mass/volume} \tag{2.1}$$

   With regard to water, for example, perhaps the most common measures of density are pounds per cubic foot (lb/ft^3) and pounds per gallon (lb/gal). However, the density of a dry substance, such as sand, lime, and soda ash, is usually expressed in pounds per cubic foot. The density of a gas, such as methane, or carbon dioxide, is usually expressed in pounds per cubic foot or it is grams per liter at standard temperature (0°C) and standard pressure (760 mm mercury).

   As shown in Table 2.1, the density of a substance like water changes slightly as the temperature of the substance changes. This happens because substances usually increase in volume (size), as they become warmer. Because

TABLE 2.1
Specific Gravity and Density of Water with Temperature Change

Temperature (°F)	Specific Weight (lb/ft³)	Density (slugs/ft³)
32	62.4	1.94
40	62.4	1.94
50	62.4	1.94
60	62.4	1.94
70	62.3	1.94
80	62.2	1.93
90	62.1	1.93
100	62.0	1.93
110	61.9	1.92
120	61.7	1.92
130	61.5	1.91
140	61.4	1.91
150	61.2	1.90
160	61.0	1.90
170	60.8	1.89
180	60.6	1.88
190	60.4	1.88
200	60.1	1.87
210	59.8	1.86

of this expansion with warming, the same weight is spread over a larger volume, so the density is lower when a substance is warm than when it is cold.

Example 2.1

Problem: 12 cm³ of an object weighs 82.2 g. What is the density of the object?
Solution:

$$\text{Density of object} = \frac{82.2\text{ g}}{12\text{ cm}^3} = 6.9\text{ g/cm}^3\text{ (rounded)}$$

Example 2.2

Problem: 110 L of a gas at STP (standard temperature and pressure) measures 145 g. Calculate the density of the gas.
Solution:

$$\text{Density of the gas} = \frac{145\text{ g}}{110\text{ L}} = 1.32\text{ g/L}$$

2. **Specific Gravity—**the weight of a substance compared to the weight of an equal volume of water—a substance having a specific gravity of 2.5 weighs two and a half times more than water. This relationship is easily seen when a cubic foot of water, which weighs 62.4 lb, is compared to a cubic foot of aluminum, which weighs 178 lb. Aluminum is 2.7 times as heavy as water.

   More specifically, specific gravity is the ratio of the density of a substance to that of a standard. The standard for solids and liquids is water with a density of 1 g/cm^3. For gases, the standard is air with the density of 1.29 g/L.

   For solids and liquids,

$$\text{Specific gravity} = \frac{\text{density of the substance}}{\text{density of water}} \tag{2.2}$$

   For gases,

$$\text{Specific gravity} = \frac{\text{density of the gas}}{\text{density of air}} \tag{2.3}$$

Example 2.3

Problem: Calculate the specific gravity of a metal with the density of 21.2 g/cm^3.
Solution:

$$\text{Specific gravity of a metal} = \frac{21.2\ \text{g/cm}^3}{1\ \text{g/cm}^3} = 21.2$$

Example 2.4

Problem: Calculate the specific gravity of a gas with the density of 1.44 g/L.
Solution:

$$\text{Specific gravity of oxygen} = \frac{1.44\ \text{g/L}}{1.29\ \text{g/L}} = 1.12$$

When dealing with units involving weight it is not that difficult to find the specific gravity of a piece of metal. All we need to do is to weigh the metal in air, then weigh it under water. Its loss of weight is the weight of an equal volume of water. To find the specific gravity, divide the weight of the metal by its loss of weight in water.

$$\text{Specific gravity} = \frac{\text{Weight of a substance}}{\text{Weight of equal volume of water}} \tag{2.4}$$

Example 2.5

Problem: Suppose a piece of metal weighs 110 lb in air and 74 lb under water. What is the specific gravity?

Solution:

1. Step 1: 110 lb subtract 74 lb = 36 lb loss of weight in water
2. Step 2:

$$\text{Specific gravity} = \frac{110 \text{ lb}}{36 \text{ lb}} = 3.1$$

✓ **Key Point**: In a calculation of specific gravity, it is *essential* that the densities be expressed in the same units.

As pointed out, the specific gravity of water is one, which is the standard, the reference for which all other substances (i.e., liquids and solids) are compared; thus, any object that has a specific gravity greater than one (1) will sink in water. Considering the total weight and volume of a ship, its specific gravity is less than one; therefore, it can float.

For the environmental practitioner, specific gravity has a number of applications. For example, it has applications when referring to a safety data sheet (SDS) for a liquid chemical that has been accidentally spilled into one (or all) of the three environmental mediums, atmosphere, water, and/or soil. If a chemical is spilled into a water body, for example, the environmental practitioner will want to know, among other things, the chemical's specific gravity to determine if the chemical will sink or float. Obviously, this information is important because emergency response procedures may be different for a contaminant that sinks rather than for one that floats.

3. **Hardness**—commonly defined as a substance's relative ability to resist scratching or indentation. Actual hardness testing involves measuring how far an "indenter" can be pressed into a given material under a known force. A substance will scratch or indent any other substance that is softer. Table 2.2 is used for comparing the hardness of mineral substances.

   Environmental practitioners should also be familiar with another definition of hardness. In water treatment, for example, *hardness* is a characteristic of water, caused primarily by dissolved calcium and magnesium compounds or ions. Regions with limestone ($CaCO_3$) and dolomite ($MgCO_3$ and $CaCO_3$) have more hardness than others. Hardness results in increased consumption of soap; that is, simply, the harder the water the greater the consumption of soap. Generally, it is the bicarbonates (HCO_3^-), nitrates (NO_3^-), Sulfates (SO_4^{2-}), and chlorides (Cl^-) of calcium and magnesium.

   Waters are classified into several types depending on the degree of hardness. Where does the degree of hardness in water come from? Well, consider rainwater. Rainwater usually contains carbon dioxide, CO_2, in solution as carbonic acid, H_2CO_3. So, what does this mean? Well, when

TABLE 2.2
Moh's Hardness Scale

Hardness (H)	Mineral
1	Talc
2	Gypsum (fingernail: H = 2.5)
3	Calcite (penny: H = 3)
4	Fluorite
5	Apatite
6	Feldspar (Glass plate: H = 5.5)
7	Quartz
8	Topaz

water percolates through deposits of limestone (consists of calcite and aragonite) and dolomite (consisting of calcium magnesium carbonate), a few of these carbonates are converted into soluble bicarbonates.

$$CaCO_3 + H_2CO_3 \rightarrow Ca(HCO_3)_2$$

Chlorides and sulfates of calcium and magnesium are also dissolved by rainwater. In addition, the presence of metals such as manganese (Mn) and iron (Fe) may also cause hardness. Water can be classified depending on the degree of hardness: as soft 15–250 mg/L (pr [[,) as $CaCO_3$, as medium hard 50–100 mg/L or ppm as $CaCO_3$, or as very hard 100–2—mg/: or ppm as $CaCO_3$.

Water hardness can cause many maintenance problems, especially with piping and process components where scale buildup can occur. Water hardness can also be a nuisance in laundering, due to the wasting of detergent and the collection of precipitate on fibers.

4. **Odor**—odors are characteristic of many substances. Some have pleasant odors, like amyl acetate (fruity, banana, fragrant, sweet, and earthy odor); some have pungent odors, like ammonia; some have disagreeable odors, like acetic acid (vinegar, sour). The question is: do rare earth elements have odors? It depends. It depends on the aggregate material they are attached to. For example, volcanic muds containing rare earth elements usually give off the odor of hydrogen sulfide (H_2S—rotten egg odor).

DID YOU KNOW?

Hard water can be softened by boiling, using lime-soda ash, and by ion exchange.

For the environmental practitioner, odor can be an important parameter or first indication of potential trouble or hazards. For example, environmental practitioners (and rare-earth miners) are often involved with confined space entry operations. By configuration (i.e., limited access, enough room for worker entry, not normally occupied, and the possibility of hazardous atmosphere) confined spaces are inherently dangerous. If before entry or during entry, the entrants detect a very distinct rotten egg odor, this should alarm the pre-entrants or entrants of the presence of hydrogen sulfide, a very toxic gas at certain concentrations. Thus, precautions as indicated in the SDS for hydrogen sulfide must be followed to protect entrants.

✓ **Caution**: Hydrogen sulfide (H_2S) at low concentration does emanate a characteristic rotten egg odor. However, continued exposure (or exposures to higher levels) to the point where its odor cannot be detected—giving those exposed a false sense of security—could be the result of numbing of the sense of smell, a very dangerous situation, obviously.

5. **Color**—color is another physical property of substances. Unless we are color blind, most of us are familiar with the colors of various substances. Pure water, for example, is usually described as colorless. Water takes on color when foreign substances such as organic matter from soils, vegetation, minerals, and aquatic organisms are present. Rare-earth metals are silvery, silvery-white, or gray in color.

   Extensive physical properties are such features as *mass*, *volume*, *length*, and *shape*. Extensive physical properties are dependent upon the amount of the substance.

CHEMICAL PROPERTIES

The non-chemist often has difficulty in distinguishing the physical versus the chemical properties of a substance. One test that can help is to ask the question: Are the properties of a substance determined without changing the identity of the substance? If we answer *yes*, then the substance is distinguished by its physical properties. If we answer *no,* then we can assume the substance is defined by its chemical properties. Simply, the *chemical properties* of a substance describe its ability to form new substances under given conditions; that is, these properties are determined by changing the chemical composition (e.g., through interaction with other substances, burning, and reaction to light and heat.)

✓ **Key Point**: The chemical properties of a substance may be considered to be a listing of all the chemical reactions of a substance and the conditions under which the reactions occur.

Another example can be used to demonstrate the difference between physical and chemical change. When a carpenter cuts pieces of wood from a larger piece of wood to build a wooden cabinet, the wood takes on a new appearance. The value of the crafted wood is increased as a result of its fresh look. This kind of change, in which the substance remains the same, but only the appearance is different, is called a *physical change*. When this same wood is consumed in a fire, however, ashes result. This change of wood into ashes is called a *chemical change. In a chemical change a new substance is produced.* Wood has the property of being able to burn. Ashes cannot burn.

KINETIC THEORY OF MATTER

There are three states of matter—*solids*, *liquids*, and *gases*. All matter is made of *molecules*. Matter is held together by attractive forces, which prevent substances from coming apart. The molecules of a solid are packed more closely together and have little freedom of motion. In liquids, molecules move with more freedom and are able to flow. The molecules of gases have the greatest degree of freedom, and their attractive forces are unable to hold them together.

The *Kinetic Theory of Matter* is a statement of how we believe atoms and molecules behave and how it relates to the ways we have to look at the things around us. Essentially, the theory states that all molecules are always moving. More specifically, the theory says:

- All matter is made of atoms, the smallest bit of each element. A particle of a gas could be an atom or a group of atoms.
- Atoms have an energy of motion that we feel as temperature. At higher temperatures, the molecules move faster.

✓ **Interesting Point**: *Kinetic* is the Greek word that means, "motion." *Theory* is the Greek word for "idea."

Matter changes its state from one form to another. Examples of how matter changes its state include

- melting is the change of a solid into its liquid state
- freezing is the change of a liquid into its solid state
- condensation is the change of a gas into its liquid form
- evaporation is the change of a liquid into its gaseous state
- sublimation is the change of a gas into its solid state and vice versa (without becoming liquid)

✓ **Key Point**: Water freezing at 32°F is an example of a physical property of water.

ATOMIC THEORY

To this point, we have described how chemical change involves a complete transformation of one substance into another. Now that we have established a basic understanding of the chemical change of substances, we need to look at the structure of matter.

The ancient Greek philosophers believed that all the matter in the universe was composed of four elements: earth, fire, air, and water. Today, we know of more than 100 different elements and scientists are attempting to create additional elements in their laboratories. An *element* is a substance from which no other material can be obtained (see Figure 2.1).

Today, we know that *atoms* are the basic building blocks of all *matter. Atoms* equal the smallest particle of matter with constant properties (see Figure 2.2).

Atoms are so small that it would take approximately TWO THOUSAND MILLION atoms side by side to equal one meter in length (see Figure 2.3)!

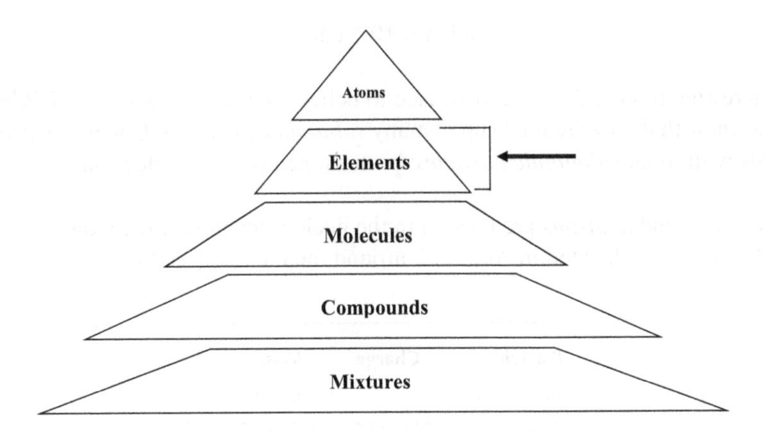

FIGURE 2.1 An element stands alone.

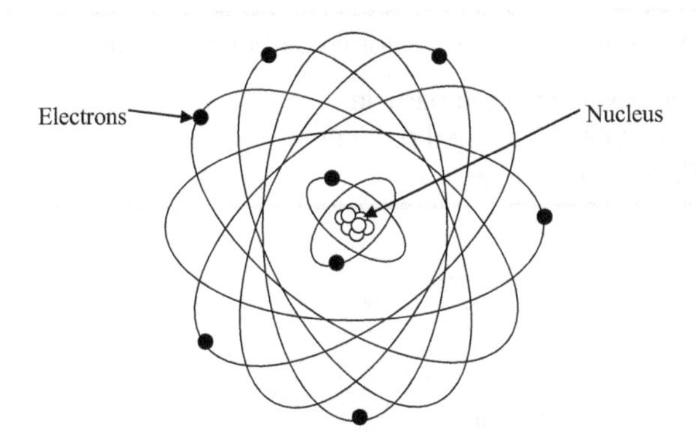

FIGURE 2.2 Atom.

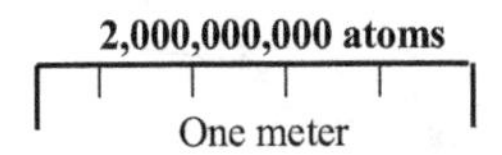

FIGURE 2.3 Two thousand million atoms side by side equal one meter in length.

Atoms are so small that scientists were forced to devise special weights and measures

- Mass/weight: atomic units (au)

$$\text{» } 1 \text{ au} = 1.6604 \times 10^{-24} \text{ g}$$

- Length: Angstrom (Å)

$$\text{» } 1 \text{ Å} = 10^{-8} \text{ cm}$$

It is interesting to note that scientists used to believe that atoms were *indivisible*, but we now know that they are made up of many *subatomic particles.* Chemistry primarily deals with three *subatomic particles*: protons, neutrons, and electrons.

- *Protons* and *neutrons* are located in the nucleus (center) of the atom.
- *Electrons* are located in "orbitals" around the nucleus (see Figure 2.4).

Particle	Charge	Mass
Proton (P)	+	1 atomic unit
Neutron (N)	No charge	1 atomic unit
Electron (e)	–	None

✓ **Key Point**: An atom contains sub-atomic particles including:

Protons (+) positively charged particle
Electrons (–) negatively charged particle
Neutrons (0) have no charge

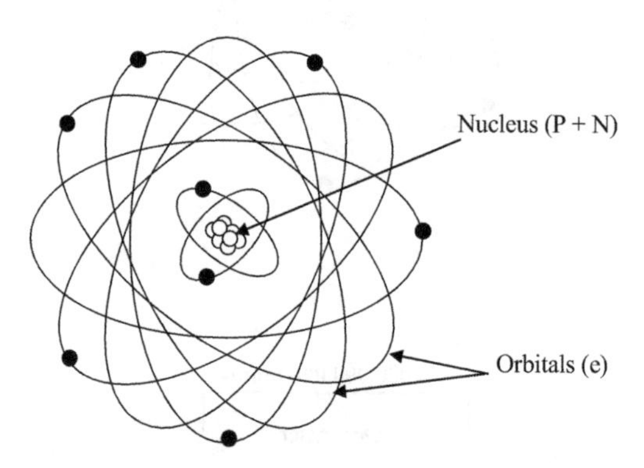

FIGURE 2.4 Orbitals.

The *stability* of a nucleus depends on the balance between:

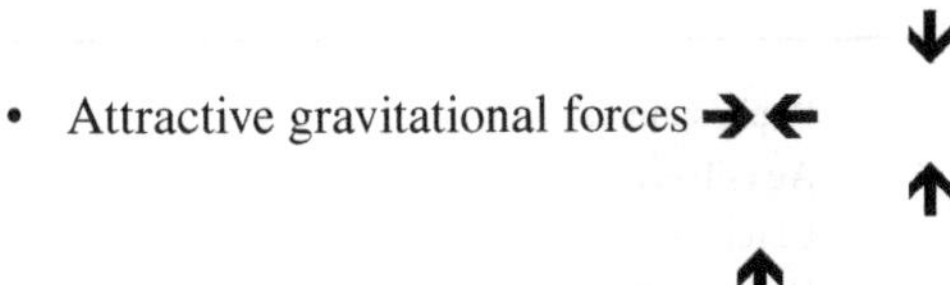

- Attractive gravitational forces

- Repulsive electronic forces
- The ratio of protons to neutrons

✓ **Key Point**: In a stable atom (or neutral atom), the number of electrons = the number of protons.

ELEMENTS

As mentioned, scientists have identified more than 100 diverse types of *atoms*, which they call *elements. An atom is the smallest unit of an element, which still retains the* properties of that element.

Of the more than 100 elements, only 83 are not naturally radioactive, and of those only 50 or so are common enough to our experience to be useful in this text. These elements, though, are going to stay the same for a long time; that is, they will outlast any political entity.

✓ **Key Point**: An *element* is a basic substance from which no other material can be obtained. Stated differently, the chemical elements are the simplest substances into which ordinary matter may be divided. All other materials have more complex structures and are formed by the combination of two or more of these elements.

Elements, of which 92 are naturally occurring and the remaining are synthesized, are substances; therefore, they have physical properties that include density, hardness, odor, color, etc., and chemical properties that describe the ability to form new substances (i.e., a list of all the chemical reactions of that material under specific conditions).

Each element is represented by a *chemical symbol.* Chemical symbols are usually derived from the element's name (e.g., Al for aluminum). The chemical symbols for the *elements* known in antiquity are taken from their Latin names (e.g., Pb for lead) but written in English. For every element, there is one and only one upper case letter (e.g., O for oxygen). There may or may not be a lower-case letter with it (e.g., Cu for copper). When written in chemical equations, we represent the elements by the symbol alone with no charge attached.

Examples of Chemical Symbols

Fe (iron)	P (phosphorus)
Al (aluminum)	Ag (silver)
Ca (calcium)	Cl (chlorine)
C (carbon)	H (copper)
N (nitrogen)	K (potassium)
Rn (radon)	He (helium)
H (hydrogen)	Si (silicon)
Cd (cadmium)	U (uranium)

✓ **Key Point**: Presently, we know only 100+ elements, but well over a million compounds. Only 88 of the 100+ elements are present in detectable amounts on Earth, and many of these 88 are rare. Only 10 elements make up approximately 99% (by mass) of the Earth's crust, including the surface layer, the atmosphere, and bodies of water (See Table 2.3).

As can be seen from Table 2.3, the most abundant element on Earth is oxygen, which is found in the free state in the atmosphere as well as in combined form with other elements in numerous minerals and ores.

Matter and Atoms

Molecules (see Figure 2.5) consist of two or more ***atoms*** that have chemically combined

$$A + B \rightarrow C$$

When *atoms* (elements) chemically combine to form *compounds*, they *lose all their original properties* and create a new set of properties, unique to the compound (see

TABLE 2.3
Elements Making 99% of Earth's Crust, Oceans, and Atmosphere

Element	Symbol	% of Composition	Atomic Number
Oxygen	O	49.5	8
Silicon	Si	25.7	14
Aluminum	Al	7.5	13
Iron	Fe	4.7	26
Calcium	Ca	3.4	20
Sodium	Na	2.6	11
Potassium	K	2.4	19
Magnesium	Mg	1.9	12
Hydrogen	H	1.9	1
Titanium	Ti	0.58	22

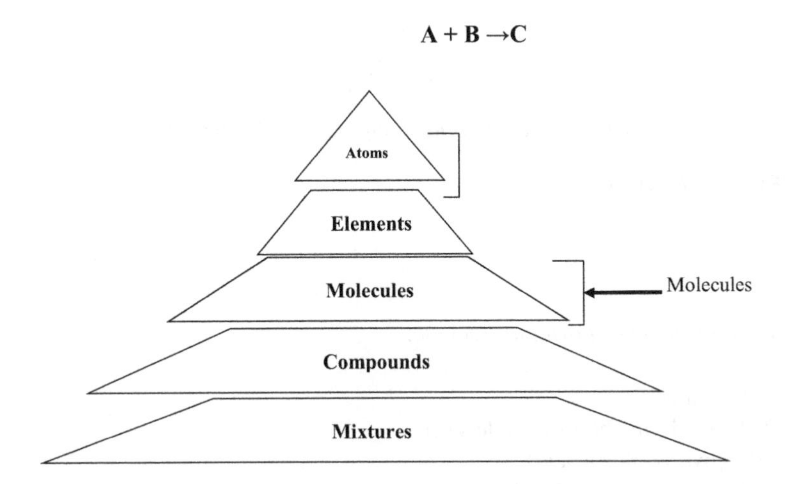

FIGURE 2.5 Molecules.

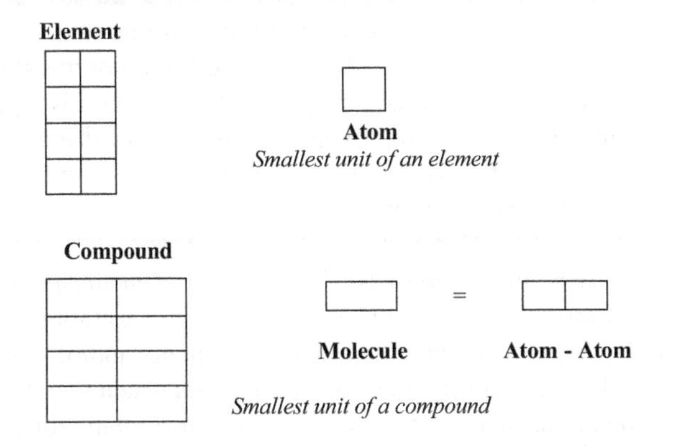

FIGURE 2.6 Compound.

Figure 2.6). For example, sodium (Na) and chlorine (Cl) are poisonous, but they combine to form a compound called sodium chloride, which is ordinary table salt.

A *molecule* is the smallest unit of a *compound*, which still retains the properties of that compound (see Figure 2.7).

Each individual *compound* will always contain *the same elements in the same proportions* by weight.

$$H_2O \neq H_2O_2$$

Water hydrogen peroxide

Compounds are represented by *chemical formulas*. Chemical formulas consist of *chemical symbols* and subscripts to describe the relative number of atoms present in each compound.

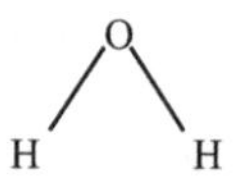

One molecule of the compound called water

FIGURE 2.7 Water molecule.

Common Chemical Formulas include:

- H_2O (water)
- NaCl (sodium chloride, table salt)
- HCl (hydrochloric acid)
- CCl_4 (carbon tetrachloride)
- CH_2Cl_2 (methylene chloride)

✓ **Key Point**: A chemical formula tells us how many atoms of each element are in the molecule of any substance. As mentioned, the chemical formula for water is H_2O. The "H" is the symbol for hydrogen. Hydrogen is a part of the water molecule. The "O" means that oxygen is also part of the water molecule. The "2" after the H means that two atoms of hydrogen are combined with one atom of oxygen in each water molecule.

Most naturally occurring matter consists of *mixtures* of *elements* and/or compounds (see Figure 2.8). Mixtures are found in rocks, the ocean, vegetation, and just about anything we find. *Mixtures* are combinations of elements and/or compounds held together by physical rather than chemical means. It is important to note that mixtures are physical combinations and compounds are chemical combinations.

Mixtures have a wide variety of compositions. Mixtures can be separated into their ingredients by physical means (e.g., filtering, sorting, distillation, etc.). The *components of a mixture* retain their own properties.

✓ **Key Point**: The thing to remember about mixtures is that we start with some pieces, combine them, and we can do something to pull those pieces apart again. We wind up with the same molecules (in the same amounts) that we started with. Two classic examples of mixtures are concrete and salt water. When mixed, they seem to form compounds, but because of physical forces (cement can still have the basic parts removed by grinding and salt water by filtering), the parts are just like they were when we started.

Compounds are two or more elements that are "stuck" (bonded) together in definite proportions by a chemical reaction. For example, water (H_2O) and halite (table

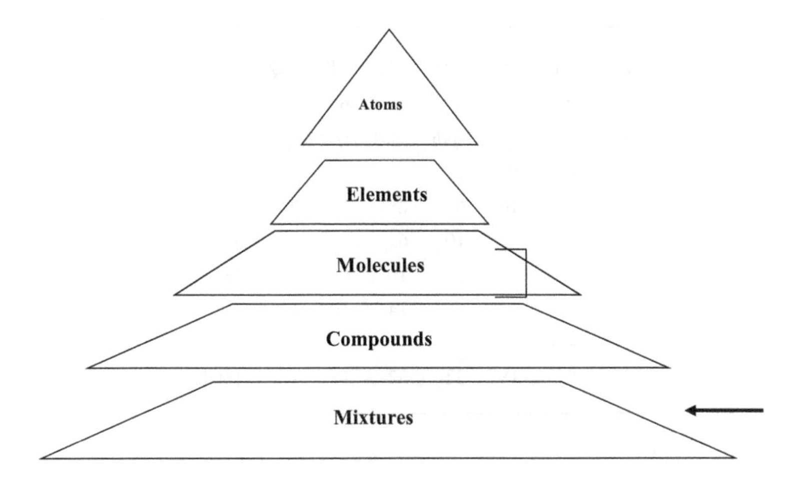

FIGURE 2.8 Mixtures of elements or compounds.

salt—NaCl). Compounds always have uniform proportions and also have unique properties that are different than their components and cannot be separated by *physical* means.

ELECTRON CONFIGURATION

Recall that an atom contains sub-atomic particles including:

- protons (+) positively charged particle
- electrons (–) negatively charged particle
- neutrons (0) have no charge

Free (unattached) uncharged atoms have the same number of electrons as protons to be electrically neutral. The protons are in the nucleus and do not change or vary except in some nuclear reactions. The electrons are in discrete pathways or shells (orbits) around the nucleus. There is a ranking or hierarchy of the shells, usually with the shells further from the nucleus having a higher energy.

In an atom, the electrons seek out the orbits that are closest to the nucleus, because they are located at a lower energy level. The low-energy orbits fill first. The higher energy levels fill with electrons only after the lower energy levels are occupied. The lowest energy orbit is labeled as the *K shell*, which is closest to the nucleus. The outer orbits, or shells, are listed in alphabetical order: *K, L, M, N, O, P, and Q.*

In the atomic diagram shown below, it can be seen that two electrons are needed to fill the K shell, eight electrons to fill the L shell, and, for light elements (atomic numbers 1–20), eight electrons will fill the M shell.

K	L	M	N	O	P	Q	
1	2	3	4	5	6	7	
s	*sp*	*spd*	*spdf*	*spdf*	*spd*	*sp*	
2	8	8	2				20
		10	6	2			38
			10	6	2		56
			14	10	6	2	88
				14	10	6	
------	------	------	------	------	------	------	
2	8	18	32	32	18	8	Totals

Electron Configuration Chart

For the following discussion, refer to the Electron Configuration Chart above. *Electron configuration* is the "shape" of the electrons around the atom, that is, which energy level (shell) and what kind of orbital it is in. The shells were historically named for the chemists who found and calculated the existence of the first (inner) shells. Their names began with "K" for the first shell, then "L," then "M," so subsequent energy levels were continued up the alphabet. The numbers one through seven have since been substituted for the letters.

The electron configuration is written out with the first (large) number as the shell number. The letter is the orbital type (*s*, *p*, *d*, or *f*). The smaller superscript number is the number of electrons in that orbital.

To use this scheme, you first must know the orbitals. A *s* orbital only has 2 electrons. A *p* orbital has six electrons. A *d* orbital has 10 electrons. An *f* orbital has 14 electrons. We can tell what type of orbital it is by the number on the chart. The only exception to that is that "8" on the chart is "2" plus "6," that is, *s* and a *p* orbital. The chart reads left-to-right and then down to the next line, just as English writing. Any element with over 20 electrons in the electrical neutral unattached atom will have all the electrons in the first row on the chart.

The totals on the right indicate using whole rows. If an element has an atomic number over 38, take all the first two rows and whatever more from the third row. For example, Iodine is number 53 on the Periodic Table (discussed later). For its electron configuration, we would use all the electrons in the first two rows and 15 more electrons: 1*s*2 2*s*2 2*p*6 3*p*6 4*s*2 3*d*10 4*p*6 5*s*2 from the first two rows and 4*d*10 5*p*5 from the third row. We can add up the totals for each shell at the bottom. Full shells would give us the totals on the bottom.

PERIODIC TABLE OF ELEMENTS

The Periodic Table of Elements (see Figure 2.9) is an arrangement of the elements into rows and columns in which those elements with comparable properties occur in the same column.

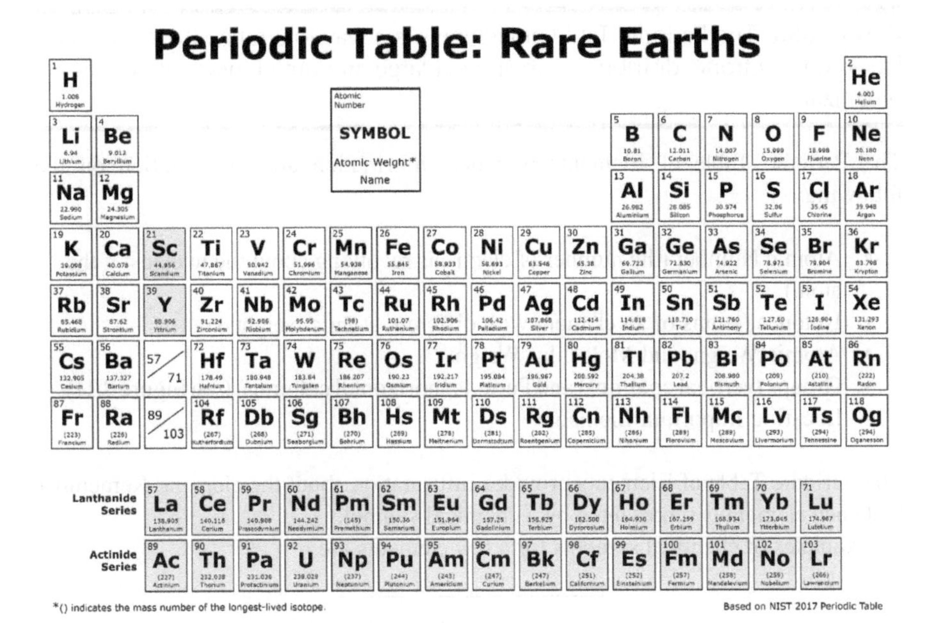

FIGURE 2.9 The Periodic Table. In this illustration and later in Figure 3.1 elements 57–71 and 21, 39 which are the focus of this book.

Source: NIST Periodic Table accessed 10/17/2021 @ http://www.NIST.gov. (U.S. Dept of Commerce).

Chemists use this table to correlate the chemical properties of known elements and predict the properties of new ones. Figure 2.10 shows the element iron and how it is charted on the periodic table.

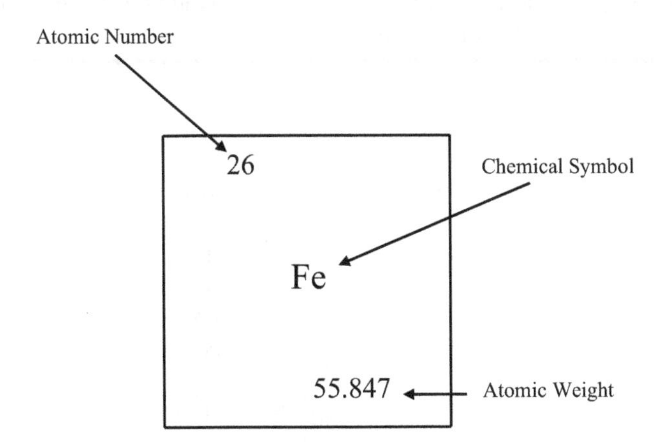

FIGURE 2.10 The element iron.

✓ **Key Point**: The Periodic Table of Elements is a way to arrange the elements, based on electronic distribution, to show a large amount of information and organization.

The Periodic Table of Elements provides information and organization on the following:

- Atomic Number (the number of electrons revolving about the nucleus of the atom)
- Isotopes (atoms of an element with different atomic weights)
- Atomic Weight and Molecular Weight
- Groups (vertical columns) and Periods (horizontal columns) (see Figure 2.11)
- Locating Important Elements

The Periodic Table of Elements provides information about the element. Remember that each element is represented by a chemical symbol.

Examples of chemical symbols for a few elements are

Fe (iron)	P (phosphorus)
Al (aluminum)	Ag (silver)
Ca (calcium)	Cl (chlorine)
C (carbon)	H (hydrogen)

Atomic Number is the number of protons in the nucleus of an atom (also the number of electrons surrounding the nucleus of an atom). Remember that *protons* are positively charged subatomic particles found in the nucleus of the atom.

✓ **Key Point**: The number of protons in the nucleus determines the atomic number. Change the number of protons and we change the element, the atomic number, and the atomic mass.

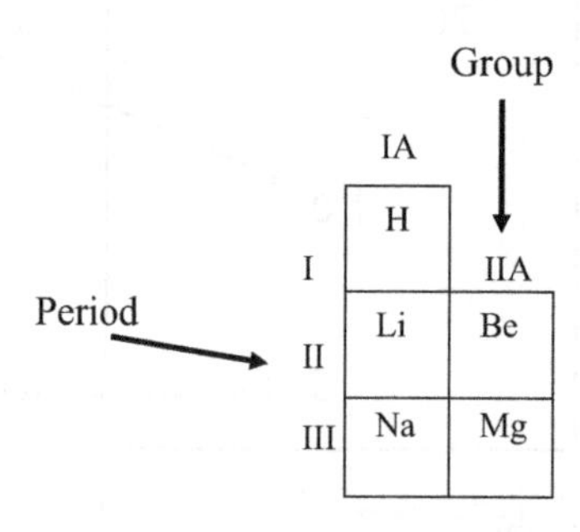

FIGURE 2.11 Periods and groups.

As mentioned, the *atomic number* can also indicate the *number of electrons* if the charge of the atom is known.

The *atomic number* (again, the number of protons) is a *unique identification number* for each element. It *cannot change* without changing the identity of the element.

$$C = 6 \quad O = 8 \quad Cl = 17$$

The *number of neutrons* can change without changing the identity of the element or its chemical and physical properties.

Atoms with the same number of protons but different numbers of neutrons are called *isotopes.*

Isotope	#P	#N
^{1}H (hydrogen)	1	0
^{2}H (deuterium)	1	1
^{3}H (tritium)	1	2

Note that if the number of neutrons changes the *atomic weight* of the atom changes.

Isotope	#P	#N	Weight
^{1}H (hydrogen)	1	0	1 au
^{2}H (deuterium)	1	1	2 au
^{3}H (tritium)	1	2	3 au

Atomic weight—the relative weight of an average atom of an element, based on C^{12} being exactly 12 atomic units. Chemists generally *round* atomic weights to the nearest whole number.

$$H = 1 \quad O = 16 \quad Fe = 56$$

Chemists add up *atomic weights* to determine the *molecular weight* of a compound. For example, the *molecular weight* of water (H_2O) equals the weight of 2 hydrogen atoms and 1 oxygen atom. Hydrogen has a molecular weight of 1.008 and oxygen has a molecular weight of 15.999. The total molecular weight then is $18.015 \cong 18$. If the mass is in grams, divide by the molecular weight. This results in the amount of moles.

Molecular weights are an important indication of the general size (and therefore complexity) of a compound. In general (but not always), the higher the molecular weight, the greater the likelihood that the compound may persist in the environment.

Periods, the rows (horizontal lines) of the periodic table, are organized by increasing atomic number. *Groups* are the *columns* of the Periodic Table, which contain elements with similar chemical properties. Within each Period (row), the *size*

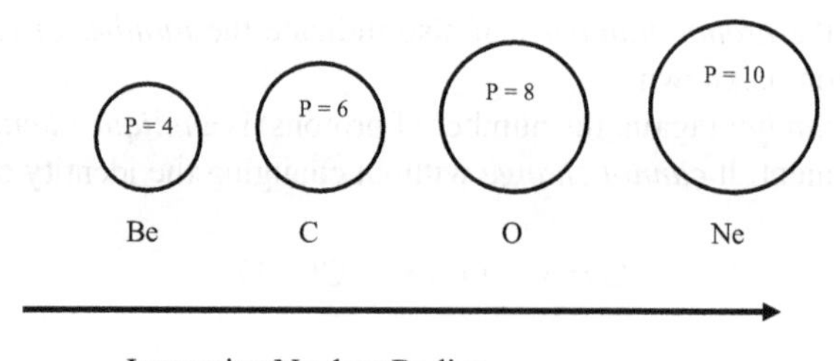

FIGURE 2.12 Nuclei increase with an increase in atomic number.

of the nuclei increases going from left to right because the *atomic number* increases (see Figure 2.12).

Within each *Group* (column), the *size* of the atoms increases going from top to bottom because the *number of electron shells* increases. Within each Group (Column), elements have similar chemical reactivity because they have a *similar number of outer shell electrons*). The Group numbers (i.e., Roman numerals) above each column indicate the number of outer shell electrons in each Group:

Group V = 5 outer shell electrons
Group II = 2 outer shell electrons
Group VII = 7 outer shell electrons

Only "outer shell" electrons are involved in chemical change. The nucleus and inner shell electrons are not altered in any way during ordinary chemical reactions.

Using the periodic table, you should also be able to locate the following:

- The alkali metals (Group I)
- The halogens (Group VII)
- The Noble Gases (Group VIII)
- Metals, nonmetals, and metalloids
- The rare earths and radioactive elements

The Alkali Metals

- Lithium (Li)
- Sodium (Na)
- Potassium (Na)
- Rubidium (Rb)
- Cesium (Cs)

The Halogens

- Fluorine (F)
- Chlorine (Cl)

- Bromine (Br)
- Iodine (I)

The Noble Gases

- Helium (He)
- Neon (Ne)
- Argon (Ar)
- Krypton (Kr)
- Xenon (Xe)
- Radon (Rn)

Metals, Non-metals, and Metalloids

Typically, a diagonal line across the right-hand side of the Periodic Table separates the ***metals*** from the **non-metals**. The **metalloids** are Boron (B), Silicon (Si), Germanium (Ge), Arsenic (As), Antimony (Sb), and Tellurium (Te).

The Rare Earths and Radioactive Elements

These elements are located in separate rows at the bottom of the Periodic Table (see Figure 2.9). The *Rare Earths* or *Lanthanides* all resemble Lanthanide in their chemical reactivity. The *Actinides* resemble Actinide reactivity, and all are radioactive.

The Periodic Table is a systematic arrangement of the elements with similar chemical reactivities and allows chemists to predict trends in reactivity.

✓ **Key Point**: The Periodic Table is a systematic arrangement of the elements, which groups elements with similar chemical reactivities and allows chemists to predict trends in reactivity.

THE BOTTOM LINE FOR REVIEW

Associate

Atomic number with: number of protons in nucleus and identity of element

Neutrons with: isotopes; two atoms of the same element are isotopes If they differ in the number of neutrons in their nuclei

Mass number with: number of protons plus number of neutrons

Atomic weight with: average of mass numbers for all isotopes of an element as they occur on the Earth

3 Rare Earth Elements

INTRODUCTION

Unless you are a chemist or have studied the periodic table in school you might not be familiar with or ever heard of rare earth elements (REEs) a group of 15 chemical elements in the periodic table (see Figure 3.1), specifically the lanthanides. Two other elements, scandium (Sc) and yttrium (Y), have a similar physiochemistry to the lanthanides, are commonly found in the same mineral assemblages, and are often referred to as REEs. REEs are, on average, more abundant than precious metals (e.g., gold, silver, and platinum), but historically, they're called "rare" because although relatively abundant in the earth's crust, they rarely occur in concentrated forms and are found in low concentrations in minerals, making them economically challenging to obtain—difficult to isolate.

With regard to chemical properties of REEs, it's all about how electrons fill their shells. The rare earths represent the 4f-electron series or configuration in the periodic table (see Figure 3.1); for example, the REE Neodymium, atomic number 60, has

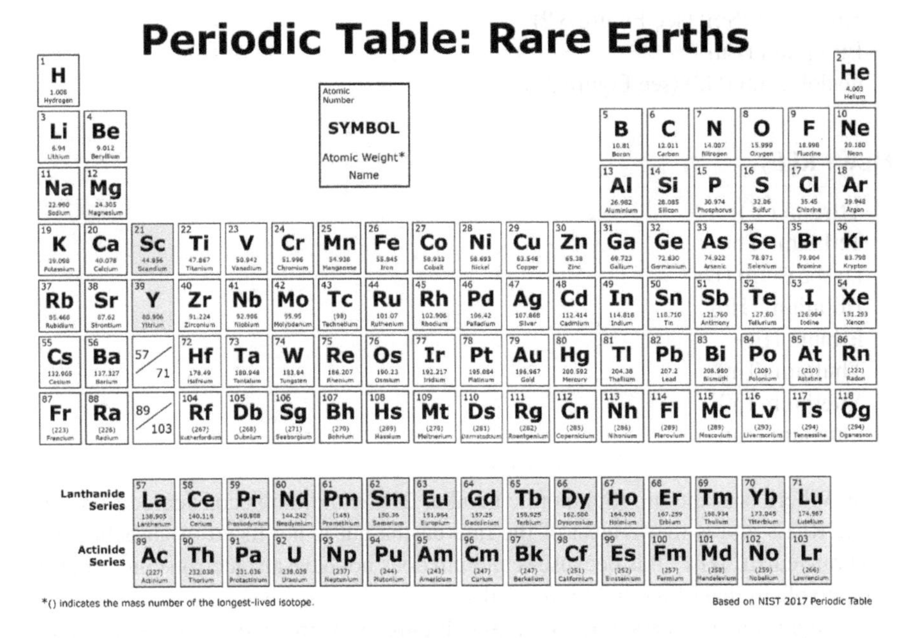

FIGURE 3.1 Periodic table with elements 57–71 and 21, 39 which are the focus of this book.

Source: NIST Periodic Table accessed 10/17/2021 @ http://www.NIST.gov. (U.S. Dept of Commerce).

DOI: 10.1201/9781003350811-4

an electron configuration of [xe]4f[4]6s[2]. These electrons are hidden in the electronic structure, allowing them to have very untainted and distinct atomistic properties. It is this configuration that gives them incredibly effective and handy magnetic and optical properties.

REEs are classified as "critical minerals" meaning they are a non-fuel or mineral materials essential to the economic and national security of the United States. They serve an essential function in the manufacturing of a product, the absence of which would have significant consequences for our economy or our national security.[1] It should be noted that their supply chain is vulnerable to disruption,

STATUS OF RARE EARTH ELEMENTS

Again, REEs are comprised of the lanthanides, scandium, and yttrium. REEs are classified as "light" and "heavy" based on atomic number. Light REEs (LREEs) consist of (atomic numbers 57 and 64):

LIGHT REEs

Lanthanum (La) (see Figure 3.2)
Cerium (Ce) (see Figure 3.2)
Praseodymium (PR) (see Figure 3.2)
Neodymium (Nd) (see Figure 3.2)
Promethium (Pm)
Samarium (Sm) (see Figure 3.2)
Europium (Eu)
Gadolinium (Gd) (see Figure 3.2)

HEAVY REEs

Yttrium (Y)
Terbium (Tb)
Dysprosium (Dy)
Holmium (Ho)
Erbium (Er)
Thulium (Tm)
Ytterbium (Yb)
Lutetium (Lu)

Note: Where is scandium (Sc) atomic number 21? Scandium is considered to be one of the REEs and is the lightest of the rare earths but is not listed or classified as a light rare earth. Why? It is because scandium does not have an electron configuration comparable to that of the light rare earths.

Neodymium and praseodymium (see Figure 3.2) are two of the key critical LREEs used in the manufacture of neodymium–iron–boron (NdFeB) magnets. NdFeB magnets have the highest energy product (magnetic strength) among commercially available magnets and make possible high energy density and high energy

FIGURE 3.2 Rare Earth Oxides. Clockwise from top center: praseodymium, cerium, lanthanum, neodymium, samarium, and gadolinium.

Source: Peggy Greb, U.S. Department of Agriculture. Agricultural Research Service.

efficiency in energy technologies.[2] To increase the operating temperature of the magnets, the HREEs dysprosium and terbium are often added to the NdFeB alloy. Note that HREEs tend to be less abundant and more expensive than LREEs. This point is important because the deployment of energy technologies such as wind turbines and electric vehicles (EVs) could lead to serious imbalances of supply and demand for these key materials.

NOTES

1 Executive Order 13817. Executive Office of the President 2017. Accessed 10/19/2021 @ https://www.whitehouse.govpresidential-actions/presidential-executive-order-federal-strategy-ensure-secure-reliable-supplies-critical-materials/.

2 *Critical Mineral Resources of the United States—Economic and Environmental Geology and Prospects for Future Supply*. U.S. Geological Survey (2017). Accessed 10/20/2021 @ https://pubs.er.usgs.gov/publication/pp1802.

4 Information Please!

THE 411 ON CHEMICALS[1]

After presenting two of the foundation blocks, basic electricity and basic chemistry terminology and definitions it is important to bring forth another foundation block dealing with REEs. This foundation block has to do with the proper way in which to introduce each of the REEs correctly. It is my opinion, based on more than 40 years of experience in the field and elsewhere, that the safety and health of anyone involved with anything is important and should never be overlooked or ignored. Therefore, this chapter deals with one of OSHA's major safety standards that is pertinent to our discussion of the 17 elements that make up the REEs. Whenever we want to find information about almost anything, there are several sources that we can use. For instance, we can go to the library and look up whatever information is available on the topic or subject we are trying to define or get more in-depth information on. We could also ask a teacher who lectures on the topic or subject. Probably the most frequently used source of information at the present time is the Internet. If you do use the internet all you have to do is to use one of the search engines and it is likely that you will receive more answers than you can possibly read or use. It helps in your search if you ask the right question—phrase it correctly and so forth.

So, what does any of this have to do with rare earth elements (REEs)? This is a good question and phrased quite correctly. Well, if we want to learn about REEs we can enroll in a chemistry class and when the lecture addresses REEs we can pay attention and if not satisfied with the information, we can ask questions. Hopefully, you will have the type of instructor who is not immune to questions and not too impatient (or, heaven forbid, not privy to the answer) to provide answers. Another source of information on REEs is a book on the subject; maybe a book like this one (good luck trying to locate one). And, again, both the library and the Internet might provide you with an answer to your enquiry—and based on my surfing the net I know you will be able to browse through more information than you can possibly need, but the 411 on your search is available.

Now, the question shifts to where is the 411—the information—that I need, and what information is best? That is, what information is best, most dependable, and in some cases (unfortunately) truly accurate?

The truth be told, information about REEs is rather scant or is in a form that many individuals will find difficult to comprehend unless the person is well-versed in science and in particular chemistry.

So, when I need information on chemicals, any chemical, or chemical compound that is accurate, concise, in plain English (or in whatever language I need) and in modern-day parlance, right on, I go to the chemical manufacturer—to the people who mine, produce, refine, and distribute the chemical.

DOI: 10.1201/9781003350811-5

Will these chemical producers actually provide the information we need? Good question. And the simple answer is yes, and the compound answer is, by and according to law, they must.

Law?

What law?

OSHA's Hazard Communication Standard (HCS) 29 CFR 1910.1200 owes its genesis to the Bhopal Incident which occurred in December of 1984. The deadly gas methyl isocyanate (MIC) was accidentally released, contaminating an untold number of people and with a death toll estimated at 8,000 initially and another 8,000 or more weeks later.

The Bhopal Incident, the ensuing chemical spill, and the resulting tragic deaths and injuries are well known. Now, however, not all of the repercussions—and the lessons learned—from this incident are not as well known. After Bhopal, there arose a worldwide outcry.

"How could such an incident occur? Why wasn't something done to protect the inhabitants? Weren't there safety measures taken, or in place, to prevent such a disaster from occurring?"

"Lots of questions, few answers. The major problem was later discovered to be a failure to communicate. That is, the workers, residents, and visitors had no idea of how dangerous a chemical spill could be. Far too many found out the hard way."

In the U.S., these questions, and others, along with "after the fact findings" were bandied around, and talked about by the President, and by Congress. Because of Bhopal, Congress took the first major step to prevent such incidents from occurring in the U.S. What Congress did was to direct OSHA to take a close look at chemical manufacturing in the U.S. to see if a Bhopal-type Incident could occur in this country. OSHA did a study and then reported to Congress that a Bhopal-type incident in the U.S. was very unlikely.

Tragically, within only a few months of OSHA's report to Congress, a chemical spill occurred, similar to Bhopal, but fortunately, not deadly (*no deaths, but 100+ people became ill*), in the town of Institute, West Virginia.

DID YOU KNOW?

Exposure to hazardous chemicals is one of the most serious dangers facing American workers today, and too many workers may not even understand the risk that they're taking when working with chemicals.

Needless to say, Congress was upset. Because of Bhopal and the Institute, West Virginia fiascoes, OSHA mandated its Hazard Communication Program, 29 CFR 1910.1200 in 1984. Later, other programs like SARA (Superfund) Title III, reporting requirements for all chemical users, producers, suppliers, and storage entities, were mandated by USEPA.

There is no "all-inclusive list" of chemicals covered by the HazCom Standard; however, the regulation refers to "any chemical which is a physical or health hazard." Those specifically deemed hazardous include:

- Chemicals regulated by OSHA in 29 CFR Part 1910, Subpart Z, Toxic and Hazardous Substances
- Chemicals included in the American Conference of Governmental Industrial Hygienists' (ACGIH) latest edition of Threshold Limit Values (TLVs) for *Chemical Substances and Physical Agents in the Work Environment.*
- Chemicals found to be suspected, or confirmed carcinogens by the National Toxicology Program, in the *Registry of Toxic Effects of Chemical Substances*, published by NIOSH, or appearing in the latest edition of the *Annual Report on Carcinogens*, or by the International Agency for Research on Cancer, in the latest editions of its IARC *Monographs.*

Congress decided that those personnel working with, or around, hazardous materials, "had a right to know" about those hazards. Thus, OSHA's HCS was created. The HCS is, without a doubt, the most important regulation in the communication of chemical hazards to employees. Moreover, because OSHA's Hazard Communication is a dynamic (*living*) standard, it has been easily amendable, and adjusted to comply with ongoing, worldwide changes in an effort to make employer and worker chemical safety compliance requirements, more pertinent, and applicable.

Considering this on-going desire for currency and applicability, Federal OSHA published a revised HCS (*HazCom*), on March 26, 2012, which aligns with the *United Nation's Globally Harmonized System of Classification and Labeling of Chemicals.*

This revision affects how chemical hazards are classified, the elements incorporated into a label, and the format of the safety data sheet (SDS). In addition, some terminology and several definitions have changed, including the definition of a hazardous chemical.

Under its HCS (*more commonly known as "HazCom" or the "Right to Know Law*"), OSHA requires employers who use, or produce chemicals on the worksite, to inform all employees of the hazards that might be involved with those chemicals. HazCom says that employees have the right to know what chemicals they are handling or may be exposed to. HazCom's intent is to make the workplace safer. Under the HazCom Standard, the employer is required to fully evaluate all chemicals on the worksite for possible physical and health hazards. All information relating to these hazards must be made available to the employee 24 hours each day. The standard is written in a performance manner, meaning that the specifics are left to the *employer* to develop.

HazCom also requires the *employer* to ensure proper labeling of each chemical, including chemicals that might be produced by a *process* (process hazards). For example, in the oil, wastewater industries, deadly methane gas is generated in the underground stream and/or waste stream. OSHA's HazCom requires the employer to label hazardous chemical hazards (and other chemicals), so that workers are warned, and safety precautions are followed.

Labels must be designed to be clearly understood by all workers. Employers are required to provide both training, and written materials, to make workers aware of what they are working with, and what hazards they might be exposed to. Employers are also required to make SDSs available to all employees. An SDS is a fact sheet for a chemical that poses a physical or health hazard in the workplace. An SDS must be in English and contain the following information:

- Identity of the chemical (label name)
- Physical hazards
- Control measures
- Health hazards
- Whether it is a carcinogen
- Emergency and first aid procedures
- Date of preparation of the latest revision
- Name, address, and telephone number of manufacturer, importer, or other responsible parties

Blank spaces are not permitted on an SDS. If relevant information in any one of the categories is unavailable at the time of preparation, the SDS must indicate that no information was available. Your facility must have an SDS for each hazardous chemical it uses. Copies must be made available to other companies working on your worksite (*outside contractors, for example*), and they must do the same for you. The facility Hazard Communication program must be in writing and, along with an SDS, be made available to all workers, 24 hours each day/each shift.

Better Communication for Worker Safety and Health

To provide better worker protection from hazardous chemicals, and to help American businesses compete in a global economy, OSHA has revised its Hazardous Communication (HazCom) standard to align with the united Nations' Globally Harmonized System of Classification and Labeling of Chemicals—referred to as *GHS*. This *GHS* incorporates the quality, consistency, and clarity of hazard information that workers receive, by providing harmonized criteria for classifying, and labeling, all hazardous chemicals, and for preparing SDSs for these chemicals.

The *GHS* system is an innovative approach that has been developed through international negotiations and embodies the knowledge gained in the field of chemical hazard communication since the HazCom standard was first introduced in 1983. Simply, *HazCom*, along with *GHS*, means better communication to reduce chemical hazards for workers on the job.

Benefits of HAZCOM with GHS[2]

Practicing Occupational Safety and Health professionals are familiar with OSHA's original 1983 HCS. Many are now becoming familiar with the phase-in of the newly combined *HazCom* and GHS standard. The first thing they learn is that the Globally Harmonized System (GHS) is an international approach to hazard communication, providing agreed-upon criteria for the classification of chemical hazards, and a standardized approach to labeling elements, and SDS. The *GHS* was negotiated in a multi-year process by hazard communication experts from many different countries, international organizations, and stakeholder groups. It is based on major existing systems around the world, including OSHA's HCS and the chemical classification, and labeling systems, of other US agencies.

The result of this negotiation process is the United Nations' document entitled "Globally Harmonized System of a Classification and Labeling of Chemicals," commonly referred to as *The Purple Book*. This document provides harmonized classification criteria for health, physical, and environmental hazards of chemicals. It also includes standardized label elements that are assigned to these hazard classes, and categories and provides the appropriate signal words, pictograms, and the hazard, and precautionary statements necessary to convey the types of hazards to users. A standardized order of information for SDSs is also provided. These recommendations can be used by regulatory authorities, such as OSHA, to establish mandatory requirements of hazard communication, but do not constitute a model regulation.

OSHA's motive to modify the *HCS* was to improve the safety and health of workers, through more effective communications on chemical hazards. Since it was first promulgated in 1983, the *HCS* has provided employers, and employees, extensive information about the chemicals in their workplaces. The original standard is performance oriented, allowing chemical manufacturers, and importers, to convey information on labels, and on SDSs, in whatever format they choose. While the available information has been helpful in improving employee safety and health, a more standardized approach to classifying the hazards, and for conveying the information, will be more effective, and provide further improvements in American workplaces. The GHS provides such a standardized approach, including detailed criteria for determination of what hazardous effects a chemical can cause, as well as standardized label elements assigned by hazard class, and category. This will enhance both employer and worker comprehension of the hazards, which will help to ensure the appropriate handling, and safe use, of workplace chemicals. In addition, the SDS requirements establish an order of information that is standardized. The harmonized format of the SDSs will enable employers, workers, health professionals, and emergency responders to access the information more efficiently, and effectively, thus increasing their utility.

Adoption of the *GHS* in the US, and around the world, will also help to improve information received from other countries—since the US is both a major importer and exporter of chemicals. American workers often see labels and SDSs from other countries. The diverse, and sometimes conflicting, national, and international requirements can create confusion among those who seek to use hazard information more effectively. For example, labels and SDSs may include symbols, and hazard statements, which are unfamiliar to readers, or may not be well understood. Containers may be labeled with such a large volume of information, *(overkill)* that important statements are not easily recognized. Given the differences in hazard classification criteria, labels may also be incorrect when used in other countries. If countries around the world adopt the *GHS*, these problems will be minimized, and chemicals crossing borders will have consistent information, thus improving communication globally.

Phase-in Period for the Hazard Communication Standard

During the phase-in period (2013–2016), employers were required to be following either the existing HCS, or the revised HCS, or both. OSHA recognized that hazard communication programs would go through a period where both labels, and SDS sheets, would be present in the workplace, under both standards. This was considered acceptable, and employers were not required to maintain two sets of labels, and SDS sheet, for compliance purposes.

It is important to point out that prior to OHSA's effective compliance date for full implementation of the revised HCS, employee training was required to be conducted. This was the case because American workplaces received the new SDS and labeling requirements before the full compliance date was met. Thus, employees needed to be trained early to enable them to recognize, and understand, the new label elements (*i.e., pictograms, hazard statements, precautionary statements, and signal words*) and the SDS format.

Major Changes to the Hazard Communication Standard

There are three major areas of change in the modified HCS: in hazard classification, labels, and SDSs.

- **Hazard Classification**: The definitions of hazard have been changed to provide specific criteria for the classification of health and physical hazards, as well as classification of mixtures. A result of these specific criteria is to ensure that evaluations of hazardous effects are consistent across manufacturers and that labels and SDSs are more accurate.
- **Labels**: Chemical manufacturers and importers will be required to provide a label that includes a harmonized signal word, pictogram, and hazard statement for each hazard class, and category. Precautionary statements must be provided.
- **Safety Data Sheets (SDS)**: Will now have a 16-section format.

Note: The GHS does not include harmonized training provisions but recognizes that training is essential to achieve an effective hazard communication approach. The revised HCS requires that workers be re-trained within two years of publication of the final result, and this would serve to facilitate more effective recognition, and understanding, of the new labels and SDSs.

HAZARD CLASSIFICATION

Not all HCS provisions are changed in the revised HCS. The revised HCS is simply a modification to the existing standard and has been designed to make it universal and worker friendly. The parts of the standard that did not relate to the GHS (*such as the basic framework, scope, and exemptions*) remained largely unchanged. There have been some modifications in terminology as an effort to align the revised HCS more closely with the language used in the GHS. For example, the term "hazard determination" has been changed to "hazard classification" and "material safety data sheet" was changed to "safety data sheet."

Under both the current HCS and the revised HCS, an evaluation of chemical hazards must be performed considering the available scientific evidence concerning such hazards. Under the current HCS, the hazard determination provisions have definitions of *hazard*, and the evaluator had to determine whether the data on a chemical met those definitions. It is a performance-oriented approach that provides parameters for the evaluation, but not specific, detailed criteria.

The hazard classification approach in the revised HCS is quite different. The revised HCS has specific criteria for each health and physical hazard, along with detailed instructions for hazard evaluation, and determinations as to whether mixtures or substances are covered. It also establishes both hazard *classes*, and hazard *categories*—for most of the possible effects caused. The classes are divided into categories that reflect the relative severity of the effect. The current HCS does not include categories for most of the health hazards covered, so this innovative approach provides additional information that can be helpful in providing the appropriate response to deal with the hazard more effectively. OSHA has included the general provisions for hazard classification in paragraph (d) of the revised rule and added extensive appendices that address the criteria for each health or physical effect.

Label Changes under the Revised HCS

Under the current HCS, the label preparer must provide the identity of the chemical, and the appropriate hazard warnings. This may be done in a variety of ways, and the method to convey the information is left to the preparer. Under the *revised HCS*, once the hazard classification is completed, the standard specifies what information is to be provided for each hazard class, and category. Labels will require the following elements:

- ***Pictogram***: a symbol plus other graphic elements, such as a border, background pattern, or color that is intended to convey specific information about the hazards of a chemical. Each pictogram consists of a specified symbol on a white background, within a red square set on a point, or diamond shape (*i.e., a white background with a border*) (see Figure 4.1). There are nine pictograms under the GHS. However, only eight pictograms are required under the HCS.
- ***Signal Words***: a signal word on the label is used to indicate the relative level of severity of a hazard and alerts the reader to a potential hazard. The signal words used are "danger" and "warning" (see Figure 4.2). "Danger" is used for the more severe hazards, while "Warning" is used for less severe hazards.
- ***Hazard Statement***: a statement assigned to a hazard class and category, which describes the nature of the hazard(s) of a chemical including, where appropriate, the degree of hazard.
- ***Precautionary Statement***: a phrase that describes recommended measures to be taken to minimize, or prevent, adverse effects resulting from exposure to a hazardous chemical or due to the improper storage or handling of a hazardous chemical.

In the revised HCS, OSHA is lifting the stay on enforcement regarding the provision to update labels when additional information on hazards becomes available.

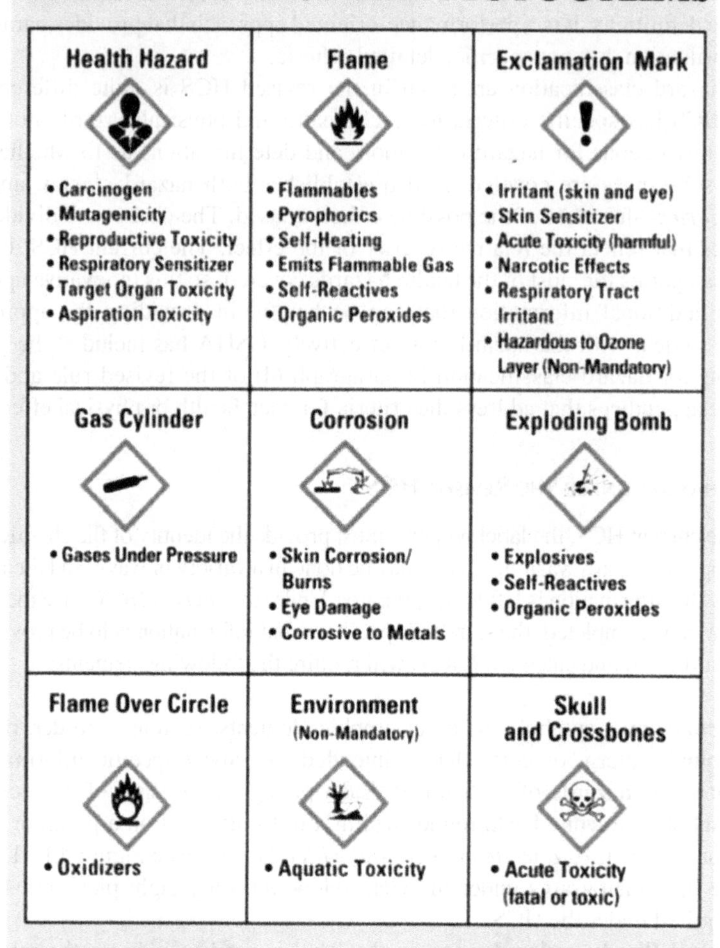

HAZCOM STANDARD PICTOGRAMS
Health Hazard
• Carcinogen
• Mutagenicity
• Reproductive Toxicity
• Respiratory Sensitizer
• Target Organ Toxicity
• Aspiration Toxicity
Flame
• Flammables
• Pyrophorics
• Self-Heating
• Emits Flammable Gas
• Self-Reactives
• Organic Peroxides
Exclamation Mark
• Irritant (skin and eye)
• Skin Sensitizer
• Acute Toxicity (harmful)
• Narcotic Effects
• Respiratory Tract Irritant
• Hazardous to Ozone Layer (Non-Mandatory)
Gas Cylinder
• Gases Under Pressure
Corrosion
• Skin Corrosion/ Burns
• Eye Damage
• Corrosive to Metals
Exploding Bomb
• Explosives
• Self-Reactives
• Organic Peroxides
Flame Over Circle
• Oxidizers
Environment
(Non-Mandatory)
• Aquatic Toxicity
Skull and Crossbones
• Acute Toxicity (fatal or toxic)

FIGURE 4.1 Global harmonized labels.

Chemical manufacturers, importers, distributors, or employers who become newly aware of any significant information regarding the hazards of a chemical, shall revise the labels for the chemical within six months of becoming aware of the latest information. If the chemical is not currently produced or imported, the chemical manufacturer, importer, distributor, or employer shall add such information to the label before the chemical is shipped or introduced into the workplace again.

The current standard provides employers with flexibility regarding the type of system to be used in their workplaces, and OSHA has retained that flexibility in the revised HCS. Employers may choose to label workplace containers either with the same label that would be on the shipped container for the chemical under the revised rule, or with label alternatives that meet the requirements for the accepted standard.

FIGURE 4.2 Sample signal word labels.

Alternative labeling systems, such as the National Fire Protection Association (NFPA) 704 Hazard Rating, and the Hazardous Material Identification System (HMIS), are permitted for workplace containers. However, the information supplied on these labels must be consistent with the revised HCS, (i.e., no conflicting hazard warnings or pictograms).

SDS Changes under the Revised HCS

The information required on the (Material) SDS will remain essentially the same as in the current standard (HazCom, 1994). HazCom 1994 indicates what information must be included on an SDS, but does not specify a format for presentation, or order of information. The revised HCS (2012) requires that the information on the SDS be presented in a specified sequence. The revised SDS should contain 16 headings (Table 4.1).

HAZCOM AND OSHA COMPLIANCE

When Rachel Hazard, an experienced federal OSHA auditor, made her walk-around of the Chemmix facility, of all the compliance items she was specifically looking for, few were more important to her than those required by Hazard Communication including a written program, employee training, SDSs, evidence of on-site hazard determination, and hazard labeling.

Rachel is well aware of the historical record. During a typical audit, OHSA auditors usually find and list at least four major violations regarding HazCom noncompliance. These four items are lacking with such frequency that they fall somewhere in the top ten "most often cited" violations. They include: (1) facility lacks a written Hazard Communication Program; (2) facility fails to label chemical hazards; (3) facility fails to train employees on HazCom; and (4) facility fails to make SDS available to all employees.

During the walk-around, Rachel is specifically looking for all four of these HazCom items. Before the walk-around even started, she knew whether or not the facility had a written HazCom program. Remember, at the arrival conference Rachel

TABLE 4.1
Minimum Information for an SDS

1. Identification of the substance, or mixture, and of the supplier	• GHS product identifier • Other means of identification • Recommended use of the chemical and restrictions on use • Supplier's details (including name, address, phone number, etc.) • Emergency phone number
2. Hazards identification	• GHS classification of the substance/mixture, and any national, or regional information • GHS label elements, including precautionary statements. (*Hazard symbols may be provided as a graphical reproduction of the symbols in black and white or the name of the symbol, e.g., flame, skull, and crossbones*) • Other hazards which do not result in classification (*e.g., dust explosion hazard*) or are not covered by GHS
3. Composition/-information on ingredients	**Substance** • Chemical identity • Common name, synonyms, etc. • CAS number, EC number, etc. • Impurities and stabilizing additives which are themselves, classified, and which contribute to the classification of the substance **Mixture** • The chemical identity and concentration, or concentration ranges, of all ingredients which are hazardous within the meaning of the GHS and are present above their cutoff levels.
4. First aid measures	• Description of necessary measures, included according to the different routes of exposure i.e., inhalation, skin and eye contact, and ingestion • Most important symptoms/effects, acute and delayed • Indication of immediate medical attention and special treatment needed, if necessary
5. Firefighting measures	• Suitable (*and unsuitable*) extinguishing methods • Specific hazards arising from the chemical (*e.g., nature of any hazardous combustion products*) • Special protective equipment and precautions for firefighters
6. Accidental release measures	• Personal precautions, protective equipment, and emergency procedures • Environmental precautions • Methods and materials for containment and cleaning up
7. Handling and storage	• Precautions for safe handling • Conditions for safe storage, including any incompatibilities
8. Exposure controls/-personal protection	• Control parameters, e.g., occupational exposure limit values, or biological limit values • Appropriate engineering controls • Individual protection measures, such as personal protective equipment
9. Physical and chemical properties	• Appearance (*physical state, color, etc.*) • Odor • Odor threshold • pH • melting point/freezing point

(*continued*)

TABLE 4.1 ***Continued***
Minimum Information for an SDS

	• initial boiling point and boiling range • flash point • evaporation rate • flammability (*solid, gas*) • upper/lower flammability, or explosive limits • vapor pressure • vapor density • relative density • solubility (*is*) • partition coefficient: n-octanol/water • autoignition temperature • decomposition temperature
10. Stability and reactivity	• Chemical stability • Possibility of hazardous reactions • Conditions to avoid (*e.g., static discharge, shock, or vibration*) • Incompatible materials • Hazardous composition products
11. Toxicological information	Concise, but complete, and comprehensible description of the various toxicological (*health*) effects and available data used to identify those effects, including: • Information on the likely routes of exposure (*inhalation, ingestion, skin, and eye contact*) • Symptoms related to the physical, chemical, and toxicological characteristics • Delayed and immediate effects and also chronic effects from short- and long-term exposure
12. Ecological information	• Ecotoxicity (aquatic and terrestrial, where available) • Persistence and degradability • Bio accumulative potential • Mobility in soil • Other adverse effects
13. Disposal considerations	• Description of waste residues and information on their safe handling, and methods of disposal, including the disposal of any contaminated packaging
14. Transportation information	• UN number • Transport Hazard class(es) • Packing group, if applicable • Marine pollutant (*Yes/No*) • **Special** precautions which a user needs to be aware of, or needs to comply with, in connection with transport; or conveyance either within, or outside that premises
15. Regulatory information	• Safety, health, and environmental regulations specific for the product in question
16. Other information including information on preparation and revision of SDS	

asked for a copy of several different safety programs, including HazCom. She hasn't had a chance to read the programs, but she will.

If we assume that Chemmix does have a written HazCom Program, Rachel will be able to make a judgment about how effective the program is, simply by observing facility conditions as she makes her rounds.

To gain some understanding of exactly what HazCom items Rachel will be looking for (the items that should be included in the written HazCom program) during her walk-around, let's take a look at what she will look for in a typical HazCom program. Remember, this is a particular or specific audit example. Programs must be tailored to meet the exact needs of individual facilities.

OSHA Hazard Communication Audit Items

- Does the facility have a written HazCom Program?
- Is the written HazCom available for the perusal of employees, outside contractors, and visitors?
- On review of the facility's written program, does the written program specify individual responsibilities for implementation and adherence to the facility's HazCom Program?
- Are employees acquainted with the facility's HazCom Program?
- Do all personnel have easy access to SDS's?
- Have all employees been trained on HazCom?
- Whenever a new physical, or health hazard, is introduced into the workplace (*one for which training has not previously been accomplished*), does the employer provide the training? Specifically, does the employer provide employee training that includes:
  1. Methods and observation that may be used to detect the presence, or release of, a hazardous chemical in the work area.
  2. The physical and health hazards of the chemicals in the work area.
  3. The measures employees can take to protect themselves from these hazards, including specific procedures the employer has implemented to protect employees from exposure to hazardous chemicals, such as appropriate work practices, emergency procedures, and what personal protective equipment is to be used.
  4. The details of the hazard communication program developed by the employer, including an explanation of the labeling system, the material SDS, and how employees can obtain and use the appropriate hazard information.

Note: As with all OSHA-required training, you must not only ensure that the training is conducted, but you must also ensure that it has been properly documented.

- With regard to labeling, does the facility train all personnel on labels and other forms of warnings that are or can be elements of HazCom? Specifically, employees must be informed that the chemical manufacturer, importer, or distributor must ensure that each container of hazardous chemicals leaving the workplace is labeled, tagged, or marked with the following information:

    - Identification of the hazardous chemical(s).
    - Appropriate hazard warnings.
    - Name and address of the chemical manufacturer, importer, or other responsible party.
- Are employees informed that the employer's responsibilities include signs, placards, process sheets, batch tickets, operating procedures, or other such written materials, in lieu of affixing labels to individual stationary process containers—if the alternative method identifies the containers to which it is applicable, and conveys the information required on the label. The written materials must be readily accessible to the employees in their work area throughout each shift.
- Does the employer know that he or she must not remove or deface existing labels on incoming containers of hazardous chemicals unless the container is immediately marked with the required information? Are employees aware of this requirement?
- Are the employees aware that labels or warnings must be legible, in English, and prominently displayed on the container, or readily available in the work area throughout each work shift?

Note: Employers with employees who speak other languages may need to add the information in the employee's native language to the material presented if the information is also presented in English.

- Are employees aware that if existing labels already convey the required information, the employee need not affix new labels?
- Is the employer aware of any significant information regarding the hazards of a chemical, that he or she must revise the labels for the chemical within three months of becoming aware of the latest information?

Note: Labels on containers of hazardous chemicals shipped after that time shall contain the additional information.

- Does the facility's HazCom Program define the terms pertinent to the standard and important to employee safety and health? Are the following terms and definitions included (example list below catalogs the minimum terms/-definitions) and facility specific?

Terms and Definitions

The Hazard Communication Program defines various terms as follows: (*These terms either appear in the company's Hazard Communication Program or are definitions appropriate to SDS*).

Chemical: any element, compound, or mixture of elements and/or compounds.
Chemical Name: the scientific designation of a chemical in accordance with the nomenclature system developed by the International Union of Pure and

Applied Chemistry (**IUPAC**) or the Chemical Abstracts Service (CAS) Rules of Nomenclature, or a name which will clearly identify the chemical for conducting a hazard evaluation.

Combustible Liquid: any liquid having a flashpoint at, or above 100°F (37.8°C), but below 200°F (93.3°C).

Common Name: any designation or identification, such as code name, code number, trade name, brand name, or generic name used to identify a chemical, other than its chemical name.

Compressed Gas: a compressed gas is:

A gas or mixture of gases in a container having an absolute pressure exceeding 40 psi at 70°F (21.1°C); or

A gas or mixture of gases, in a container having an absolute pressure exceeding 104 psi at 130°F (54.4°C) regardless of the pressure at 70°F (21.1°C); or

A liquid having a vapor pressure exceeding 10 psi at 100°F (37.8°C), as determined by ASTM D-323–72.

Container: any bag, barrel, bottle, box, can, cylinder, drum, reaction vessel, storage tank, or the like, which contains a hazardous chemical.

Explosive: a chemical that causes a sudden, almost instantaneous release of pressure, gas, and heat when subjected to sudden shock, pressure, or elevated temperature.

Exposure: the actual, or potential subjection of an employee to a hazardous chemical through any route of entry, during employment.

Flammable Aerosol: an aerosol that, when evaluated by the method described in 16 CFR 1500.45, yields a flame projection exceeding 18 inches at full valve opening, or a flashback (*flame extending back to the valve*) at any degree of valve opening.

Flammable Gas: a gas that at ambient temperature, and pressure forms, a flammable mixture with air at a concentration of 13% by volume or less, or a gas that at ambient temperature, and pressure, forms a range of flammable mixtures with air wider than 12% by volume regardless of the lower limit.

Flammable Liquid: a liquid having a flashpoint of 100°F (37.8°C).

Flammable Solid: a solid, other than a blasting agent or explosive, as defined in 29 CFR 1910.109(a), that is likely to cause fire through friction, absorption of moisture, spontaneous chemical change, or retained heat from manufacturing, or processing, or which can be ignited, and that when ignited, burns so vigorously and persistently, as to create a serious hazard. A chemical is considered to be a flammable solid if, when evaluated by the method described in 16 CFR 1500.44, it ignites and burns with a self-sustained flame, at a rate greater than one-tenth of an inch per second along its major axis.

Flashpoint: the minimum temperature at which a liquid gives off a vapor in sufficient concentration to ignite.

Hazard Warning: any words, pictures, symbols, or a combination thereof appearing on a label or other appropriate form of warning which conveys the hazards of the chemical(s) in the container.

Hazardous Chemical: any chemical which is a health or physical hazard.

Hazardous Chemical Inventory List: an inventory list of all hazardous chemicals used at the site and containing the date of each chemical's SDS insertion.

Health Hazard: a chemical for which there is statistically considerable evidence based on at least one study conducted in accordance with established scientific principles that acute or chronic health effects may occur in exposed employees.

Immediate Use: the use under the control of the person who transfers the hazardous chemical from a labeled container, and only within the work shift in which it is transferred.

Label: any written, printed, or graphic material displayed on, or affixed to, containers or hazardous chemicals.

Safety Data Sheet: the written or printed material concerning a hazardous chemical, developed in accordance with 29 CFR 1910.

Mixture: any combination of two or more chemicals if the combination is not, in whole or in part, the result of a chemical reaction.

NFPA Hazardous Chemical Label: a color-code labeling system developed by the National Fire Protection Association (NFPA) that rates the severity of the health hazard, fire hazard, reactivity hazard, and special hazard of the chemical.

Organic Peroxide: an organic compound that contains the bivalent 0-0 structure, and which may be considered to be a structural derivative of hydrogen peroxide, where one or both an organic radical has replaced hydrogen atoms.

Oxidizer: a chemical (other than a blasting agent or explosive as defined in 29 CFR 1910.198(a)), that initiates, or promotes combustion in other materials, thereby causing fire, either of itself, or through the release of oxygen or other gases.

Physical Hazard: a chemical for which there is scientifically valid evidence that it is a combustible liquid, a compressed gas explosive, flammable, an organic peroxide, an oxidizer, pyrophoric, unstable, (*reactive*), or water reactive.

Portable Container: a storage vessel that is mobile, such as a drum, side-mounted tank, tank truck, or vehicle fuel tank.

Primary Route of Entry: the primary means (*by inhalation, ingestion, skin contact, etc.*) whereby an employee is subjected to a hazardous chemical.

"Right To Know" Workstation: provides employees with an essential information workstation where they can have access to site SDS sheets, Hazardous Chemical Inventory List, and Company's written Hazard Communication Program.

"Right To Know" Station Binder: A Station Binder located in the "Right To Know" workstation that contains Company's Hazard Communication Program, the Hazardous Chemicals Inventory List and corresponding SDS, and the Hazard Communication Program Review, and Signature Form.

Pyrophoric: a chemical that will ignite spontaneously in air at a temperature of 130°F (54.4°C) or below.

Signal Word: a word used on the label, to indicate the relative level of severity of a hazard, and to alert the reader to a potential hazard. The signal words used in this section are "danger" and "warning." "Danger" is used for the more severe hazards, while "Warning" is used for the less severe.

Stationary Container: a permanently mounted chemical storage tank.

Unstable (Reactive Chemical): a chemical which in its pure state, or when produced or transported, will vigorously polymerize, decompose, condense, or will become self-reactive under changing conditions of shock, pressure, or temperature.

Water reactive (Chemical): a chemical that reacts with water to release a gas that is either flammable or presents a health hazard.

Work Center: any convenient or logical grouping of designated unit processes or related maintenance actions.

RETURN TO HAZCOM AUDIT ITEMS

Because noncompliance with OSHA's Hazard Communication is the number one citable item during a routine audit, let's look at some (and some again) of the HazCom items OSHA will be looking at. You must be able to answer "yes" to each of the following items if site applicable.

- Are all chemical containers marked with the contents' names and hazards?
- Are storage cabinets used to hold flammable liquids labeled "Flammable—Keep Fire Away?"
- For a fixed extinguishing system, is a sign posted warning of the hazards presented by the extinguishing medium?
- Are all Above-ground Storage Tanks properly labeled?
- Are all hazardous chemicals which you store in any type of container, including above-ground tanks, storage tanks (*including gasoline*), or other containers holding hazardous materials, appropriately labeled with the chemical name and hazard warning?
- Are all chemicals used in spray painting operations correctly labeled?
- If you store any type of chemicals, are all containers properly labeled with chemical name, and hazard warning?

Along with checking these items, the OSHA auditor will make notes on the chemicals he or she finds in the workplace. During the walk-around, the auditor is likely to seek out any flammable materials storage lockers you have in your workplace. The auditor will list many of the items stored in the lockers. Later, when the walk-around is completed, the auditor will ask you to provide a copy of the SDS for each chemical in his or her notes.

To avoid a citation, you must not fail this major test. If the auditor, for example, noticed during the walk-around, that employees were using some type of solvent, or cleaning agent, in the performance of their work, the auditor will want to see a copy of the SDS for that particular chemical. If you can't produce a copy, you are in violation, and will be cited. Be careful on this item—it is one of the most commonly cited offenses. Obviously, the only solution to this problem is to ensure that your facility has an SDS for each chemical used, stored, or produced and that your Chemical Inventory List is current and accurate. Save yourself a big hassle—ensure that an SDS sheet is available to employees, for *each* chemical used on site.

Keep in mind that the OSHA auditor will look at each work center within your company and that each different work center will present its own specialized requirements. If your company has an environmental laboratory, for example, the auditor will spend considerable time in the lab, ensuring you are following OSHA's Laboratory Standard, and that you have a written Chemical Hygiene Plan.

THE BOTTOM LINE

Again, you may be wondering why I chose to present OSHA's Hazard Communication Program and use its SDS information as my means of describing the 17 REEs. As an environmental practitioner I have learned through experience that whenever something, anything, is found to be beneficial for our renewable energy needs along with protecting the environment, I have also found that technology of any sort is usually a double-edged sword—it has a very sharp good side and may have a very dull bad side. Nothing put forward to better our environment comes without cost. The cost may be financial or otherwise, but the point is it has a cost that is not always beneficial. Consider, for example, wind power using wind turbines to generate clean energy—a definite advantage over fossil fuels used to accomplish the same product, the same outcome. However, wind turbines are not without their negatives which include the killing of birds, the generation of noise, the unsightly appearance (depending on your point of view, of course) and the fact they are usually good for 20 years and then need to be replaced; thus, they can be costly. The question is, are the costs worth the outcome? Now I deliberately chose wind power for this example because wind turbines, especially offshore wind turbines, use permanent magnets in their generators and the permanent magnets utilize REEs simply because they are the best. So, the best outweighs the negatives—but never forget that there are negatives, and many will be pointed out later in the text.

In the following chapters, each of the REEs is described according to their public domain SDS's. Their uses are described afterward.

NOTES

1 Much of the information in this chapter is from F.R. Spellman's 2021 *Surviving an OSHA Audit*, 2nd edition. Boca Raton, FL: CRC Press.

2 Based on information from OSHA's (2014). *Modification of the Hazardous Communication Standard (HCS) to Conform with the United Nations' (UN) Globally Harmonized System of Classification and Labeling of Chemicals (GHS).* Accessed 01/-16/19 @ https://www.osha.gov/dsg/hazcom/hazcom-fag.html.

REFERENCES

HAZCOM (1994). Accessed 12/12/21 @ https://www.osha.gov/laws-regs./federalregistor.1994.
HCS (2012). OSHA hazard communication 2012 standard. Accessed 12/12/21 @ https://law.osha.gov/hazcom/2012/standard.

5 The 17

NEODYMIUM

Details

Figure 5.1 illustrates neodymium's chemical symbol, element name, atomic number, and atomic weight; Figure 5.2 illustrates neodymium's electron configuration.

- melting point—1294 K (1021°C or 1870°F)
- boiling point—3347 K (3074°C or 5565°F)
- density—7.01 g/cm^3
- solid at room temperature
- element classification—metal
- periodic element—6
- group name—lanthanide
- electronegativity—Pauling scale: 1.14

Note: In the simplest terms I can render, *electronegativity* is a measure of the ability of an atom to attract the electrons when the atom is part of a compound. When going from left to right across the periodic table electronegativity values generally

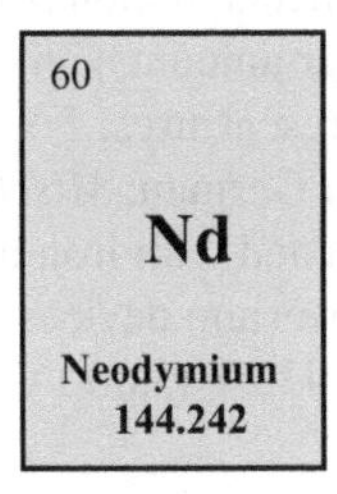

FIGURE 5.1 Chemical element designation for neodymium with chemical symbol, element name, atomic number, and atomic weight.

$1s^2$
$2s^2\ 2p^6$
$3s^2\ 3p^6\ 3d^{10}$
$4s^2\ 4p^6\ 4d^{10}\ 4f^4$
$5s^2\ 5p^6$
$6s^2$

FIGURE 5.2 Neodymium electron shell configuration.

DOI: 10.1201/9781003350811-6

increase. However, when going from top to bottom of a group, electronegativity values decrease. The highest electronegativity value is for fluorine at 4.0 and the lowest is for cesium at 0.79. And as stated above the electronegativity value for neodymium is 1.14 on the Pauling scale.

- natural occurrence—primordial (i.e., from the beginning of time)
- crystal structure—double hexagonal close-packed
- name—from the Greek *neos* and *didymos*, meaning "new twin"
- pronounced—nee-eh-DIM-ee-em

DID YOU KNOW?

Nd glass is produced by the inclusion of neodymium oxide (Nd_2O_3) in the glass melt. When this is accomplished and the glass is exposed to daylight or incandescent light the glass appears lavender in color, but under fluorescent light neodymium glass appears pale blue. The color scheme can be changed or altered to appear ranging from pure violet to wine-red or warm gray.

Neodymium is a powerful magnetic material. An Austrian chemist, Carl E. Auer von Welsbach discovered neodymium in 1885. From a material known as didymium, he separated both neodymium and the element praseodymium. At the present time, neodymium, primarily a reddish-brown phosphate mineral occurring usually in small, isolated crystals is primarily obtained via an ion exchange process monazite sand [(Ce, La, Th, Nd, Y)PO_4], a material rich in rare earth elements; lanthanum (La) is the most common rare-earth elements in monazite. Monazite is radioactive due to the presence of thorium and in lesser amounts, on occasion, uranium.

Neodymium has a variety of present uses. For example, approximately 18% of Misch metal (aka Mischmetal from German: *Mischmetall*—"mixed metal") is neodymium that when mixed and hardened with iron oxide and magnesium oxide make ignition devices in pyrophoric ferrocerium devices—misnamed "flint" but is used to produce hot sparks that can reach 5,430°F (3,000°C) used as strikers for gas welding, cutting torches and so forth.

Another important use of neodymium beginning in 1927 and still popular today is their usage as glass dyes. Neodymium compounds often are of a reddish to purple color, however, it changes with the type of lighting exposure because it contains sharp light absorption bands intermixed with emission bands of mercury and other elements. When glasses are mixed or doped with neodymium they can be used, and are used, in infrared lasers with wavelengths ranging from 1047 to 1062 nm. These are used in experiments in inertial confinement fusion (ICF)—initiates nuclear fusion by compressing and heating targets filled with thermonuclear fuel. Neodymium has an unusually large specific heat capacity (i.e., the heat required to raise the temperature of a unit mass of substance by a given amount) at liquid-helium temperatures, so it is useful in cryocoolers (a refrigerator designed to operate at 123 K, which equals −150°C or −238°F).

Neodymium is also used as a component in alloys used to make high-strength neodymium magnets. At the present time, neodymium permanent magnet (PM) generators are the trend. The widespread practice today is to use rare earth magnets in wind turbines, especially in offshore wind turbines, as they allow for high-power density and diminished size (low mass) and relatively low weight with peak efficiency at all speeds, offering a high annual production of energy with low lifetime expenditures. Most direct-drive turbines are equipped with PM generators that typically contain neodymium and smaller quantities of dysprosium. Although on a different extent or scale, the same is true for numerous gearbox designs. Using a straightforward structure, the PMs are installed on the rotor to generate a constant magnetic field. The produced electricity is collected from the stator by using the commutator, slip rings, or brushes. To lower cost, the PMs are integrated into a cylindrical cast aluminum rotor. Note that for onshore wind turbine installations, it is not necessary to utilize PM generators because the reduced size and weight are not a concern (as it is with offshore installations). PM generators are similar in operation to synchronous generators except that PM generators can be operated asynchronously (i.e., induction generators that require the stator to be magnetized from the electric grid before it works). Note that the high-power-versus-weight off-shore wind turbine generators are also used in electric motors for hybrid cars and generators for aircraft.[1] The strong, versatile permanents are widely used in high-performance hobby items, professional loudspeakers, in-ear headphones, and computer hard disks—all of these items require high power, strong magnetic fields where low mass is desired.

DID YOU KNOW?

Neodymium has been used to promote plant growth in China where rare earth elements are frequently used as fertilizer (Wei and Zhou, 1999).

Figure 5.3 shows a larger neodymium magnet attracting smaller neodymium-iron-born refrigerator magnets and a few other friends. You really can't appreciate how strong these PMs are until you attempt to pull them apart—no easy task.

Neodymium PMs used as drive motors in some vehicles require about 2.2 lb (1 kg) of neodymium per vehicle (Gorman, 2009). It is interesting to note that researchers using high-precision imaging at Radboud and Uppsala Universities observed self-induced spin glass—a behavior in the crystal structure of neodymium due to incredibly small changes in the magnetic structure (Kamber et al., 2020; Radboud University, 2020).

Personal Safety Precautions and Handling: Neodymium[2]

HMIS Ratings: Health: 1 Flammability: 1 Physical: 1
NFPA Ratings: Health: 1 Flammability: 1 Reactivity: 1

At the present time, no exposure limit has been established for neodymium. However, safety data sheets (SDSs) recommend that elemental neodymium should be handled

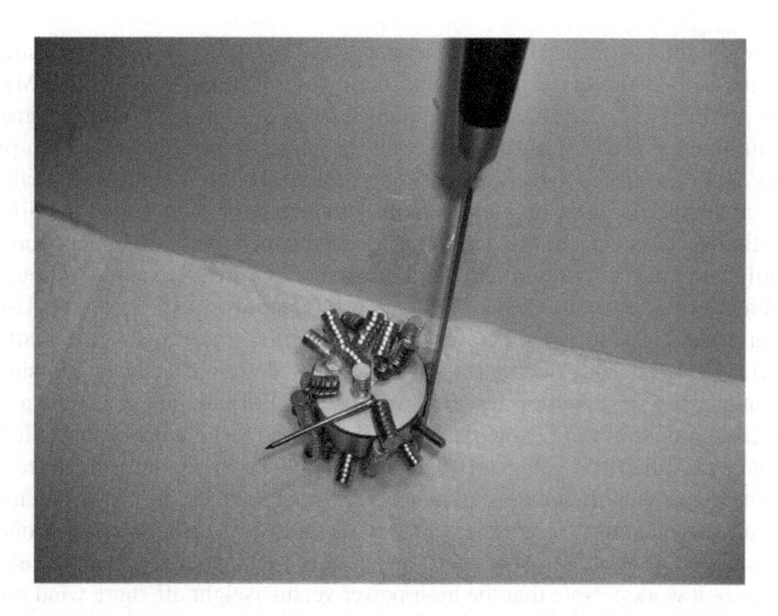

FIGURE 5.3 Neodymium permanent magnet in the process of attracting friends. Photo by F. Spellman.

in a humidity-controlled atmosphere. Handle in an enclosed, controlled process under dry argon when possible. Ensure adequate ventilation to maintain low exposure levels. Wash thoroughly before eating or smoking and do not use tobacco or food in the work area. Do not use compressed air to blow the dust off clothing or skin.

Recommended Handling Procedures for Nd

Storage

Neodymium will slowly oxidize at room temperature in the air. It should be stored under 10 torr [i.e., 1 torr = 1 mm mercury (0°C) mmHg] or better vacuum or in sealed jar under vacuum. For long-term storage, the best method is to seal this metal in evacuated Pyrex tubes with the ends fused by fusion. Oils should not be used.

Cleaning

Even when stored as described above, some surface oxidation will occur. The oxidation should be removed by filing. A wire-brush wheel may be used, but filing is preferred. If the surface has almost turned white, at least 1 mm of metal should be removed to ensure the removal of intergranular corrosion products which are near the surface. After filing, the cold worked surface can be removed by electropolishing (see below) which also passivates the surface.

Electropolishing

An electrolyte of 1% (or up to 6%) perchloric acid in absolute methanol is stirred and cooled continuously in a dry ice-acetone bath. A platinum cylinder (cup) serves as the cathode. A current density of about 0.5 amps/cm^2 usually is required. A variable voltage supply should be used and the amperage controlled to give small bubbles at the surface of the sample. The electrolyte should not be allowed to bubble excessively. The sample should be rinsed while cold in the dry-ice acetone bath, then rinsed with copious quantities of methanol.

Cutting

A metal saw (hack saw or jeweler's saw), or a low-speed diamond saw, or a spark cutter may be used. The metal should be electroplated after cutting since the freshly cut surface is quite reactive. Shearing is not recommended unless the sheared surface is filed off. The low-speed diamond saw or the spark cutter is recommended as the best method for obtaining a strain-free surface.

Cold Working

Neodymium can be cold swaged (i.e., spliced) or rolled about 30% reduction in cross-section without heat treatment. To prevent contamination, it should be wrapped or (even better) sealed in tantalum.

Handling

Because neodymium reacts primarily with moisture, it should not be touched with bare hands. Plastic gloves are recommended. It can be handled in air, but an oxide layer does form quite quickly. This layer can be removed and the surface passivated by electropolishing.

Stress Relief

The surface should be freshly cleaned by electropolishing just prior to heat treatment. A vacuum of 10 torr or better is required to prevent contamination. Minimal contamination will occur at 10 torr if the samples are wrapped in tantalum. The recommended temperature is half of the melting point in K for about 8 hours.

Melting

Neodymium may be arc or electron beam melted. Levitation and induction hearting in outgassed tantalum or tungsten crucibles are also suitable. If neodymium is heated in tantalum or tungsten to temperatures significantly above their melting points, tantalum and tungsten will dissolve in the molten rare earth.

SAMARIUM

Details

Figure 5.4 illustrates samarium's chemical symbol, element name, atomic number, and atomic weight; Figure 5.5 shows the elemental samarium; Figure 5.6 illustrates samarium's electron configuration.

Melting point—1345 K (1072°C, 1962°F)
Phase at standard temperature and pressure (STP)—solid
Appearance—silvery white

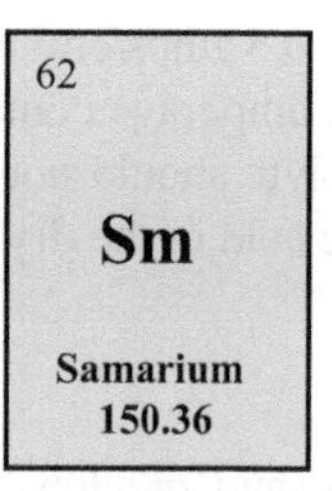

FIGURE 5.4 Chemical element designation for samarium with chemical symbol, element name, atomic number, and atomic weight.

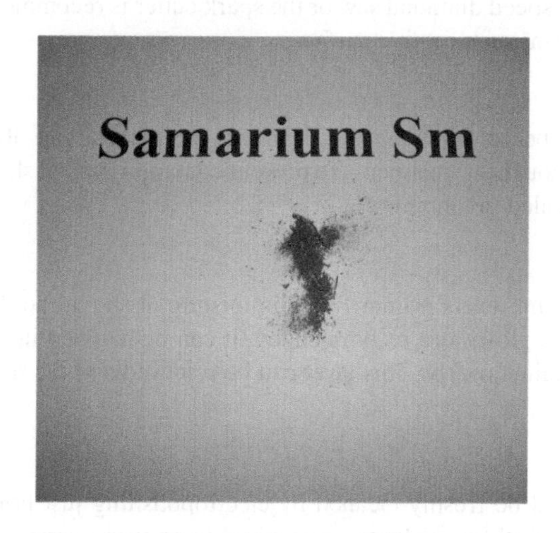

FIGURE 5.5 Elemental samarium. Photo by F. Spellman.

$$\text{Xe } 4f^6\, 6s^2$$

FIGURE 5.6 Samarium electron shell configuration.

Boiling point—1794°C
Density—7.529 g/cm^3
Solubility in water—insoluble
Heat of fusion—8.62 kJ/mol
Hardness—similar to zinc
Heat of vaporization—192 kJ/mol
Electronegativity—1.17
Natural occurrence—primordial
Magnetic ordering—paramagnetic

Note: Paramagnetic materials are those that are attracted by magnetic fields but do not retain the magnetism. In contrast, diamagnetic materials are not at all attracted by magnetic fields.

Naming—after the mineral samarskite

DID YOU KNOW?

Samarium is the hardest and the most brittle of the rare earth elements. It tarnishes in air and will ignite in air at about 150°C.

Samarium or Sm, discovered in 1879, is a lanthanide rare earth element, usually assuming the oxidation state +3. Like other elements in this group, it is a shiny metal under ordinary conditions. Samarium is only slightly toxic and has no significant biological role. Note that even though samarium is classified as a rare earth element it is the 40th most abundant element in the Earth's crust, making it more common than tin, for example. China is by far the world leader in samarium mining and production, but the element is also found in the United States, India, Sri Lanka, Australia, and Brazil.

Second, only to neodymium magnets, samarium–cobalt magnets have found wide commercial application. An advantage of samarium magnets over neodymium magnets is that samarium magnets can withstand significantly higher temperatures greater than 700°C (1,292°F), without losing their magnetic properties—this is due to samarium–cobalt's higher Curie point (i.e., the Curie temperature above which certain materials lose their permanent magnetic properties).

DID YOU KNOW?

Samarium radio isotopes are used to kill cancer cells in the treatment of prostate, lung, and breast cancer. Other applications include radioactive dating, X-ray lasers, and catalysis of chemical reactions.

Because of its high permanent magnetization—about 10,000 more times than that of iron—samarium–cobalt magnets are second only to neodymium magnets. Again, however, samarium-based magnets have higher resistance to demagnetization. Samarium magnets are used in small headphones, small motors, and high-end musical instruments.

Samarian and its compounds are also used as catalysts and chemical reagents. The catalysts assist the decomposition of plastics, dechlorination of pollutants such as PCBs (polychlorinated biphenyls), and the dehydration and dehydrogenation of ethanol (Hammond, 2004). $Sm(CF_3SO_3)_3$ (i.e., Samarian(III) triflate) is a Lewis acid catalyst—a metal-based acid that acts as an electron pair acceptor to increase reactivity of a substrate—for halogen-promoted Friedel–Crafts reaction with alkenes—attaches substituents to an aromatic ring (Hajir et al., 2007). In its oxidized form, samarium is added to ceramics and glass where it increases the absorption of infrared light. A small amount of samarium is added to mischmetal used in flint ignition devices for torches and lighters (Emsley, 2001; Hammond, 2004).

DID YOU KNOW?

Samarium is not normally part of the human diet, basically because the element is not absorbed by plants in a measurable amount. There are a few plants, however, vegetables mainly, that may contain up to 1 ppm of samarium. Samarium's insoluble salts are non-toxic, and the soluble ones are only slightly so (Emsley, 2001).

Personal Safety Precautions and Handling: Samarium[3]

HMIS Ratings: Health: 1 Flammability: 1 Physical: 1
NFPA Ratings: Health: 1 Flammability: 1 Reactivity: 1

Usually classified as a category 1 Flammable solid.

Currently, samarium's OSHA permissible exposure level (PEL) has not been established. Established engineering controls related to exposure and personal protection state that samarium should be in a humidity-controlled atmosphere and in an enclosed, controlled process under dry argon where possible. Ensure adequate ventilation to reduce exposure levels. Do not use food or drink in work area. Wash thoroughly before eating or smoking. Do not blow the dust off clothing or skin with compressed air.

Recommended Handling Procedures for Sm

Storage

Samarium will slowly oxidize at room temperature in air. It should be stored under 10 torr [i.e., 1 torr = 1 mm mercury (0°C) mmHg] or better vacuum or in sealed jar under an inert gas. For long-term storage, the best method is to seal this metal in evacuated Pyrex tubes with the ends fused by fusion. Oils should not be used.

Cleaning

If surface oxidation has occurred due to exposure to acid fumes or slightly elevated temperatures, the major portion should be removed by filing and the final polishing electrolytically (see below).

Electropolishing

For Sm, an electrolyte of 1% (or up to 6%) perchloric acid in absolute methanol is stirred and cooled continuously in a dry ice-acetone bath. A platinum cylinder (cup) serves as the cathode. A current density of about 0.5 amps/cm^2 usually is required. A variable voltage supply should be used and the amperage controlled to give small bubbles at the surface of the sample. The electrolyte should not be allowed to bubble excessively. The sample should be rinsed while cold in the dry-ice acetone bath, then rinsed with copious quantities of methanol.

Cutting

A metal saw (hack saw or jeweler's saw), or a low-speed diamond saw, or a spark cutter may be used. The metal should be electroplated after cutting since the freshly cut surface is quite reactive. Shearing is not recommended unless the sheared surface is filed off. The low-speed diamond saw or the spark cutter is recommended as the best method for obtaining a strain-free surface.

Cold Working

Sm can be cold swaged (i.e., spliced) or rolled about 10% reduction in cross section without heat treatment. To prevent contamination, it should be wrapped or (even better) sealed in tantalum.

Handling

Because samarium reacts primarily with moisture, it should not be touched with bare hands, especially if they are to be heated. Plastic gloves are recommended. It can be handled in air, but an oxide layer does form quite quickly. This layer can be removed and the surface passivated by electropolishing.

Stress Relief

The surface should be freshly cleaned by electropolishing just prior to heat treatment. A vacuum of 10 torr or better is required to prevent contamination. Minimal contamination will occur at 10 torr if the samples are wrapped in tantalum. The recommended temperature is half of the melting point in K for about 8 hours.

Melting

Sm may be arc or electron beam melted. Induction heating in outgassed tantalum or tungsten crucibles is most suitable. If samarium is heated in tantalum or tungsten to temperatures significantly above its melting point, tantalum and tungsten will dissolve in the molten rare earth.

EUROPIUM

Details

Figure 5.7 illustrates europium's chemical symbol, element name, atomic number, and atomic weight; Figure 5.8 illustrates europium's electron configuration; Figures 5.9 and 5.10 are a photo of europium.

Melting point—1099 K (826°C, 1519°F)
Phase at standard temperature and pressure (STP)—solid
Appearance—silvery white, with a pale-yellow tint; usually has oxide discoloration
Boiling point—1529°C
Specific gravity—5.264 g/cm^3
Solubility in water—decomposes

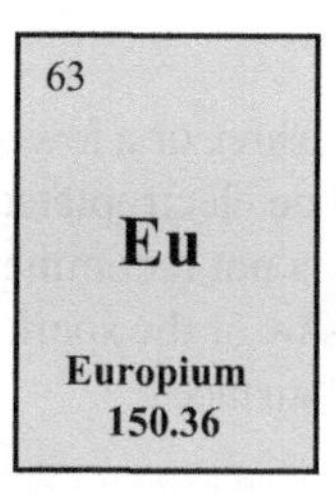

FIGURE 5.7 Chemical element designation for Europium with chemical symbol, element name, atomic number, and atomic weight.

Xe $4f^7 6s^2$

FIGURE 5.8 Europium electron shell configuration.

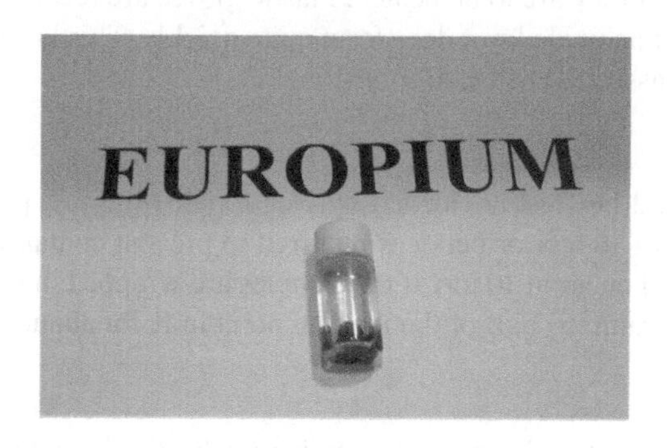

FIGURE 5.9 Europium. Photo by F. Spellman.

Heat of fusion—9.21 kJ/mol
Vickers hardness—165–200 MPa
Heat of vaporization—176 kJ/mol
Electronegativity—1.2
Natural occurrence—primordial
Magnetic ordering—paramagnetic
Naming—after Europe

Europium, Eu, is one of the rarest of the rare-earth metals. It was isolated in 1901 and is the most reactive lanthanide. It has to be stored under an inert fluid to protect it from moisture or atmospheric oxygen (Stwertka, 1996). Not found in nature as a free element, europium is the softest lanthanide; it can be dented with a fingernail and easily cut with a knife. Of all the rare-earth elements, europium is the most reactive and readily and rapidly oxidizes in air. A shiny-white metal is visible when oxidation is removed. Compared to other heavy metals, Europium is relatively non-toxic and has no significant biological role. With regard to mining, europium is mined together with other rare-earth elements, separation occurs during later processing. In use,

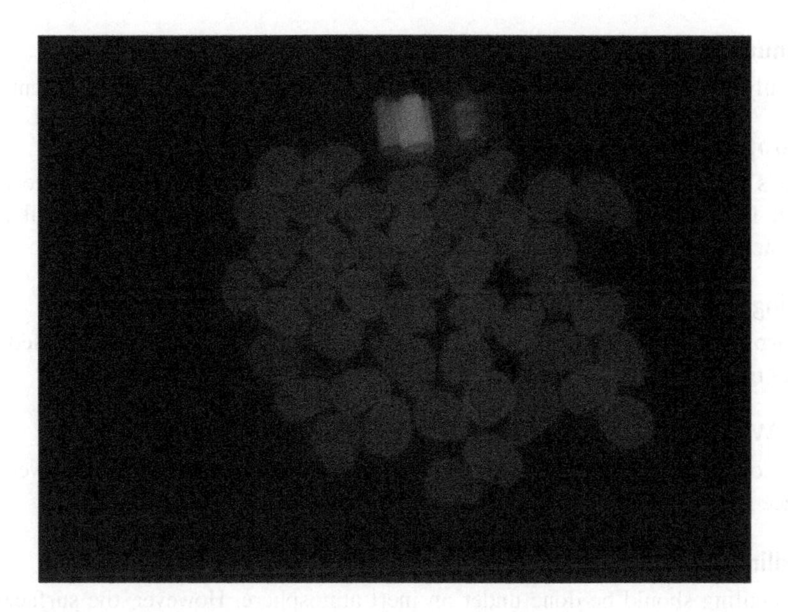

FIGURE 5.10 Europium phosphorescent powder in flasks and treated stones glowing in the dark. Photo by F. Spellman.

phosphorescent powders are often europium based. Europium is also used in liquid crystal displays, fluorescent lighting, and red and blue phosphors.

Personal Safety Precautions and Handling: Europium[4]

HMIS Ratings: Health: 1 Flammability: 3 Physical: 2
NFPA Ratings: Health: 1 Flammability: 3 Reactivity: 2 Special hazard: W

Currently, europium's OSHA PEL has not been established. Established engineering controls related to exposure and personal protection state that europium should be in a humidity-controlled atmosphere and in an enclosed, controlled process under dry argon where possible. Ensure adequate ventilation to reduce exposure levels. Do not use food or drink in the work area. Wash thoroughly before eating or smoking. Do not blow dust off clothing or skin with compressed air. Because the metal dust presents a fire and explosion hazard prepare for the possibility of a fire. Keep extinguishing agents, tools for handling, and protective clothing readily available.

Recommended Handling Procedures for Eu

Storage

Europium reacts quickly with moist air. It should be stored in Pyrex tubes or under 10 torr [i.e., 1 torr = 1 mm mercury (0°C) mmHg] vacuum or better. Sealing in a Pyrex jar under an inert atmosphere is all right for short-term storage.

Cleaning

The only method recommended is scraping the surface with a sharp instrument.

Electropolishing, Metallography

Little is known about electropolishing Eu or preparing Eu for metallographic examination. If this is necessary it is recommended that procedures developed for alkali or alkaline earth metals may work for Eu.

Cutting

A sharp knife or razorblade works the best for small pieces. A hack may be used, but blades must be changed frequently.

Cold Working

Eu is soft and can be extruded, swaged (i.e., spliced), or cold rolled. However, the surface must be protected from the air at all times.

Handling

All handling should be done under an inert atmosphere. However, the surfaces of extruded Eu react slowly in air and short time exposure to air does not contaminate the sample excessively.

Stress Relief

Must be heated under 10 torr or sealed in clean, outgassed tantalum containers. A temperature of 275–300°C for about 8 hours is recommended.

Melting

Europium should be melted in sealed, outgassed tantalum crucibles. Open crucibles under an inert atmosphere will work.

GADOLINIUM

Details

Figure 5.11 illustrates gadolinium's chemical symbol, element name, atomic number, and atomic weight; Figure 5.12 illustrates gadolinium's electron configuration; Figure 5.13 is a photo of gadolinium (Figure 5.14).

Melting point—1585 K (1312°C, 2394°F)
Phase at standard temperature and pressure (STP)—solid
Appearance—silver gray metallic
Boiling point—3000°C
Density—7.9 g/cm^3
Solubility in water—decomposes
Heat of fusion—10.05 kJ/mol
Vickers hardness—510–950 MPa
Heat of vaporization—301.3 kJ/mol

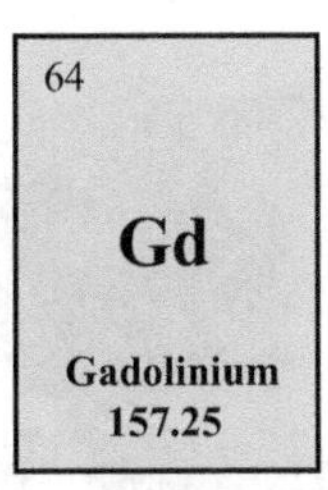

FIGURE 5.11 Chemical element designation for gadolinium with chemical symbol, element name, atomic number, and atomic weight.

Xe $4f^7$ $5d^1$ $6s^2$

FIGURE 5.12 Gadolinium electron shell configuration.

FIGURE 5.13 Gadolinium 1 mm cube. Photo by F. Spellman.

Electronegativity—1.20
Natural occurrence—primordial
Magnetic ordering—ferromagnet/paramagnetic
Naming—after the mineral gadolinite
Curie point—20°C (68°F)

Gadolinium, Gd, first isolated in 1886 is a silvery-white rare-earth metal when oxidation is removed; the oxidation occurs when the element is exposed to atmospheric oxygen and/or moisture turning its silvery-white color black, and it is found in nature only in this oxidized form. Gadolinium is produced both from monazite and bastnasite. The element is too reactive to exist naturally and is a constituent of many oxides. Gadolinium below its Curie point is ferromagnetic; its magnetic attraction rate is higher than that of nickel. However, above its Curie point, it is the most paramagnetic metal. Because it is similar to other rare-earth minerals it usually contains impurities of the other minerals when separated. Because gadolinium possesses uncommon metallurgical properties, it only takes a very slight amount of the element to enhance the workability and resistance to oxidation at elevated temperatures of chromium,

FIGURE 5.14 Gadolinium.

iron, and other similar metals. Also, because the metal or salt form of gadolinium absorbs neutrons, it is sometimes employed for shielding purposes in neutron radiography and in nuclear reactors.

Gadolinium is currently used as a magnetic resonance imaging contrast agent, in nuclear reactor shields, compact discs, and memory chips.

Personal Safety Precautions and Handling: Gadolinium[5]

HMIS Ratings: Health: 0 Flammability: 1 Physical: 1
NFPA Ratings: Health: 1 Flammability: 1 Reactivity: 1

Currently, gadolinium's OSHA PEL has not been established. Established engineering controls related to exposure and personal protection state that gadolinium should be in a humidity-controlled atmosphere and in an enclosed, controlled process under dry argon where possible. Ensure adequate ventilation to reduce exposure levels. Do not use food or drink in the work area. Wash thoroughly before eating or smoking. Do not blow dust off clothing or skin with compressed air. Because the metal dust presents a fire and explosion hazard prepare for the possibility of a fire. Keep extinguishing agents, tools for handling, and protective clothing readily available.

Recommended Handling Procedures for Gadolinium

Storage

Gadolinium does not oxidize at room temperature in air. However, for long-term storage, closed containers are recommended. Oils should not be used.

Cleaning

If surface oxidation has occurred due to exposure to acid fumes or slightly elevated temperatures, the major portion should be removed by filing and the final polishing electrolytically (see below).

Electropolishing

For Gd, an electrolyte of 1% (or up to 6%) perchloric acid in absolute methanol is stirred and cooled continuously in a dry ice-acetone bath. A platinum cylinder (cup) serves as the cathode. A current density of about 0.5 amps/cm^2 usually is required. A variable voltage supply should be used and the amperage controlled to give small bubbles at the surface of the sample. The electrolyte should not be allowed to bubble excessively. The sample should be rinsed while cold in the dry-ice acetone bath, then rinsed with copious quantities of methanol.

Cutting

A metal saw (hack saw or jeweler's saw), or a low-speed diamond saw, or a spark cutter may be used. The metal should be electroplated after cutting since the freshly cut surface is quite reactive. Shearing is not recommended unless the sheared surface is filed off. The low-speed diamond saw or the spark cutter is recommended as the best method for obtaining a strain-free surface.

Cold Working

Gd can be cold swaged (i.e., spliced) or rolled about 30% reduction in cross-section without heat treatment. To prevent contamination, it should be wrapped or (even better) sealed in tantalum.

Handling

Because gadolinium reacts primarily with moisture, it should not be touched with bare hands, especially if they are to be heated. Plastic gloves are recommended. It can be handled in the air. However, as mentioned above, strained surfaces from cutting or filing should be removed by electropolishing.

Stress Relief

The surface should be freshly cleaned by electropolishing just prior to heat treatment. A vacuum of 10 torr or better is required to prevent contamination. Minimal contamination will occur at 10 torr if the samples are wrapped in tantalum. The recommended temperature is half of the melting point in K for about 8 hours.

Melting

Gd may be arc or electron beam melted. Induction heating in outgassed tantalum or tungsten crucibles is most suitable. If lutetium is heated in tantalum or tungsten to temperatures significantly above its melting point, tantalum and tungsten will dissolve in the molten rare earth.

TERBIUM

Details

Figure 5.15 illustrates gadolinium's chemical symbol, element name, atomic number, and atomic weight; Figure 5.16 illustrates gadolinium's electron configuration.

Figures 5.17 and 5.18 show samples of terbium.

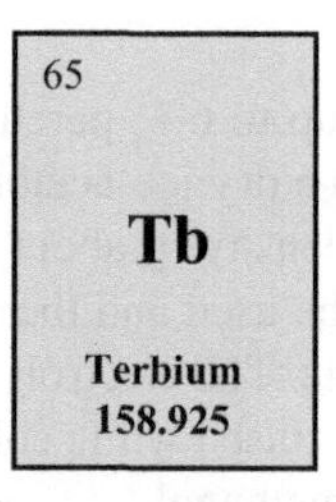

FIGURE 5.15 Chemical element designation for terbium with chemical symbol, element name, atomic number, and atomic weight.

Xe $4f^9 6s^2$

FIGURE 5.16 Terbium electron shell configuration.

FIGURE 5.17 Terbium.

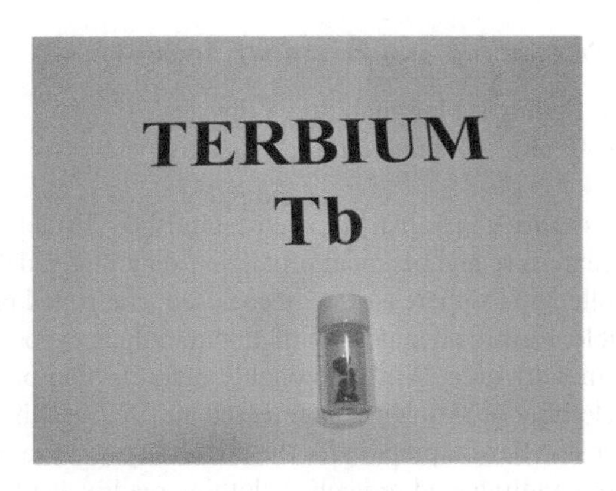

FIGURE 5.18 Terbium element. Photo by F. Spellman.

Melting point—1629 K (1356°C, 2473°F)
Phase at standard temperature and pressure (STP)—solid
Appearance—silvery white
Boiling point—3396°C
Density—8.23 g/cm^3
Solubility in water—insoluble
Heat of fusion—10.15 kJ/mol
Vickers hardness—510–950 MPa
Heat of vaporization—391 kJ/mol
Electronegativity—1.2
Natural occurrence—primordial
Magnetic ordering—paramagnetic at 300 K
Naming—after Ytterby (Sweden), where it was mined
Curie point—222 K

Terbium, Tb, first isolated in 1843 is a silvery-white rare-earth metal and is the ninth member of the lanthanide series. It is malleable, ductile, and soft enough to be easily cut with a knife. It reacts with water and is fairly electropositive. Terbium is not a free element found in nature but is found in many minerals, including cerite, gadolinite, euxenite, xenotime, and monazite. Terbium is used as a crystal stabilizer of fuel cells that operate at elevated temperatures and also used to dope calcium tungstate, calcium fluoride, and strontium molybdate. As a key component of Terfenol-D that is named after terbium, iron (Fe), and dysprosium ($Tb_xDy_{1-x}Fe_{2-x} = 0.3$) the alloy has the highest magnetostriction of any alloy, up to 0.002 m/m at saturation—it expands and contracts in a magnetic field, making it valuable and essential in the use of actuators, in naval sonar systems and in sensors. Terbium oxide is used in green phosphors and fluorescent lamps, cathode ray tubes, optical computer memories, and television tubes. When combined with divalent europium blue phosphors, and trivalent europium red phosphors, terbium green phosphors provide high-efficiency white light in trichromatic lighting technology for standard illumination in indoor lighting.

Personal Safety Precautions and Handling: Terbium

HMIS Ratings: Health: 0 Flammability: 1 Physical: 1
NFPA Ratings: Health: 1 Flammability: 1 Reactivity: 1

Currently, terbium's OSHA PEL has not been established. Established engineering controls related to exposure and personal protection state that gadolinium should be in humidity-controlled atmosphere and in an enclosed, controlled process under dry argon where possible. Ensure adequate ventilation to reduce exposure levels. Do not use food or drink in work area. Wash thoroughly before eating or smoking. Do not blow the dust off clothing or skin with compressed air. Because the metal dust presents a fire and explosion hazard prepare for the possibility of a fire. Keep extinguishing agents, tools for handling, and protective clothing readily available. Terbium has no known biological role (Hammond, 2005).

Recommended Handling Procedures for Terbium

Storage

Terbium does not oxidize at room temperature in air. However, for long-term storage, closed containers are recommended. Oils should not be used.

Cleaning

If surface oxidation has occurred due to exposure to acid fumes or slightly elevated temperatures, the major portion should be removed by filing and the final polishing electrolytically (see below).

Electropolishing

For Tb, an electrolyte of 1% (or up to 6%) perchloric acid in absolute methanol is stirred and cooled continuously in a dry ice-acetone bath. A platinum cylinder (cup) serves as the cathode. A current density of about 0.5 amps/cm^2 usually is required. A variable voltage supply should be used and the amperage controlled to give small bubbles at the surface of the sample. The electrolyte should not be allowed to bubble excessively. The sample should be rinsed while cold in the dry-ice acetone bath, then rinsed with copious quantities of methanol.

Cutting

A metal saw (hack saw or jeweler's saw), or a low-speed diamond saw, or a spark cutter may be used. The metal should be electroplated after cutting since the freshly cut surface is quite reactive. Shearing is not recommended unless the sheared surface is filed off. The low-speed diamond saw or the spark cutter is recommended as the best method for obtaining a strain-free surface.

Cold Working

Tb can be cold swaged (i.e., spliced) or rolled about 30% reduction in cross-section without heat treatment. To prevent contamination, it should be wrapped or (even better) sealed in tantalum.

Handling

Because terbium reacts primarily with moisture, it should not be touched with bare hands, especially if they are to be heated. Plastic gloves are recommended. It can be handled in the air. However, as mentioned above, strained surfaces from cutting or filing should be removed by electropolishing.

Stress Relief

The surface should be freshly cleaned by electropolishing just prior to heat treatment. A vacuum of 10 torr or better is required to prevent contamination. Minimal contamination will occur at 10 torr if the samples are wrapped in tantalum. The recommended temperature is half of the melting point in K for about 8 hours.

Melting

Tb may be arc or electron beam melted. Induction heating in outgassed tantalum or tungsten crucibles is most suitable. If terbium is heated in tantalum or tungsten to temperatures significantly above its melting point, tantalum and tungsten will dissolve in the molten rare earth.

DYSPROSIUM

Details

Figure 5.19 illustrates dysprosium's chemical symbol, element name, atomic number, and atomic weight; Figure 5.20 illustrates dysprosium's electron configuration.

Figures 5.21 and 5.22 show samples of dysprosium.

Melting point—1680 K (1407°C, 2565°F)
Phase at standard temperature and pressure (STP)—solid
Appearance—silvery white
Boiling point—2562°C
Density—8.540 g/cm^3

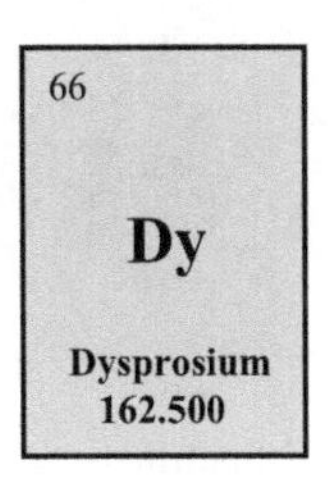

FIGURE 5.19 Chemical element designation for dysprosium with chemical symbol, element name, atomic number, and atomic weight.

Xe $4f^{10}6s^2$

FIGURE 5.20 Dysprosium electron shell configuration.

FIGURE 5.21 Dysprosium.

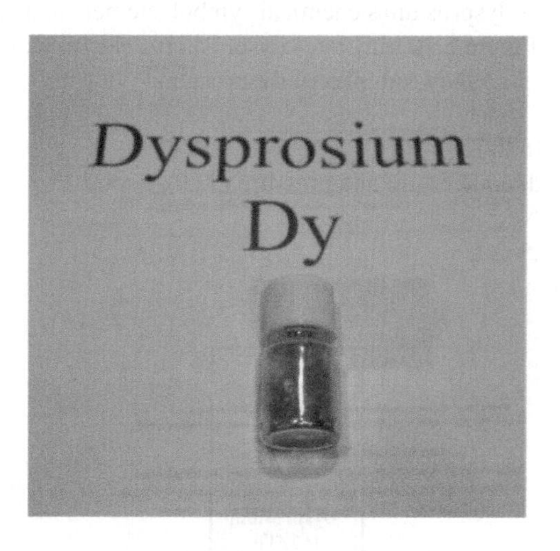

FIGURE 5.22 Dysprosium. Photo by F. Spellman.

Solubility in water—insoluble
Heat of fusion—11.06 kJ/mol
Vickers hardness—410–550 MPa
Heat of vaporization—280 kJ/mol
Electronegativity—1.22

Natural occurrence—primordial
Magnetic ordering—paramagnetic at 300 K
Naming—after the Greek, meaning "hard to get"
Curie point—87 K

Dysprosium, Dy, with a metallic silver luster was first isolated in the 1950s after the development of ion-exchange technique; it is never found in nature as a free element but instead it is found in the mineral xenotime (a rare-earth phosphate mineral), and in other minerals. Naturally occurring dysprosium is composed of isotopes, the most abundant is ^{164}Dy. In usage, dysprosium has very few uses or applications because it can be replaced by other chemical elements. When used it functions as a high thermal neutron absorber in control rods for nuclear reactors or in data-storage applications because it has high magnetic susceptibility, it is also used as a component of Terfenol-D (it expands and contracts in a magnetic field). With regard to toxicity, Dy is mildly toxic when in soluble form but not toxic in insoluble salt form. Caution is advised whenever Dy is in powder form because it may present an explosion hazard. A Dy fire cannot be extinguished with water because it reacts with the water to produce flammable gas (Dierks, 2003).

Personal Safety Precautions and Handling: Dysprosium[6]

HMIS Ratings: Health: 0 Flammability: 1 Physical: 1
NFPA Ratings: Health: 1 Flammability: 1 Reactivity: 1

Currently, dysprosium's OSHA PEL has not been established. Established engineering controls related to exposure and personal protection state that dysprosium should be within a humidity-controlled atmosphere and in an enclosed, controlled process under dry argon where possible. Ensure adequate ventilation to reduce exposure levels. Do not use food or drink in work area. Wash thoroughly before eating or smoking. Do not blow dust off clothing or skin with compressed air. Because the metal dust presents a fire and explosion hazard prepare for the possibility of a fire. Keep extinguishing agents, tools for handling, and protective clothing readily available. Use only Class D dry powder extinguishing agent.

Recommended Handling Procedures for Dysprosium

Storage

Dysprosium does not oxidize at room temperature in air. However, for long-term storage, closed containers are recommended. Oils should not be used.

Cleaning

If surface oxidation has occurred due to exposure to acid fumes or slightly elevated temperatures, the major portion should be removed by filing and the final polishing electrolytically (see below).

Electropolishing

For Dy, an electrolyte of 1% (or up to 6%) perchloric acid in absolute methanol is stirred and cooled continuously in a dry ice-acetone bath. A platinum cylinder (cup) serves as the cathode. A current density of about 0.5 amps/cm^2 usually is required. A variable voltage supply should be used and the amperage controlled to give small bubbles at the surface of the sample. The electrolyte should not be allowed to bubble excessively. The sample should be rinsed while cold in the dry-ice acetone bath, then rinsed with copious quantities of methanol.

Cutting

A metal saw (hack saw or jeweler's saw), or a low-speed diamond saw, or a spark cutter may be used. The metal should be electroplated after cutting since the freshly cut surface is quite reactive. Shearing is not recommended unless the sheared surface is filed off. The low-speed diamond saw or the spark cutter is recommended as the best method for obtaining a strain-free surface.

Cold Working

Dy can be cold swaged (i.e., spliced) or rolled about 30% reduction in cross-section without heat treatment. To prevent contamination, it should be wrapped or (even better) sealed in tantalum.

Handling

Because dysprosium reacts primarily with moisture, it should not be touched with bare hands, especially if they are to be heated. Plastic gloves are recommended. It can be handled in air. However, as mentioned above, strained surfaces from cutting or filing should be removed by electropolishing.

Stress Relief

The surface should be freshly cleaned by electropolishing just prior to heat treatment. A vacuum of 10 torr or better is required to prevent contamination. Minimal contamination will occur at 10 torr if the samples are wrapped in tantalum. The recommended temperature is half of the melting point in K for about 8 hours.

Melting

Dy may be arc or electron beam melted. Induction heating in outgassed tantalum or tungsten crucibles is most suitable. If dysprosium is heated in tantalum or tungsten to temperatures significantly above its melting point, tantalum and tungsten will dissolve in the molten rare earth.

HOLMIUM

Details

Figure 5.23 illustrates dysprosium's chemical symbol, element name, atomic number, and atomic weight; Figure 5.24 illustrates dysprosium's electron configuration.

Figures 5.25 and 5.26 show samples of holmium.

FIGURE 5.23 Chemical element designation for holmium with chemical symbol, element name, atomic number, and atomic weight.

$$Xe\ 4f^{11}6s^{2}$$

FIGURE 5.24 Holmium electron shell configuration.

FIGURE 5.25 Holmium.

Melting point—1734 K (1461°C, 2662°F)
Phase at standard temperature and pressure (STP)—solid
Appearance—silvery white
Boiling point—2600°C
Density—8.79 g/cm^3

FIGURE 5.26 Holmium. Photo by F. Spellman.

Solubility in water—insoluble
Heat of fusion—17.0 kJ/mol
Vickers hardness—410–600 MPa
Heat of vaporization—251 kJ/mol
Electronegativity—1.23
Natural occurrence—primordial
Magnetic ordering—paramagnetic
Naming—after the Latin, from *Holmia*, the Latin name for Stockholm
Curie point—20 K

Holmium, HO, is relatively soft and malleable with a metallic silver luster that was first isolated in 1878. After extraction and isolation from monazite using ion-exchange techniques it is relatively stable at room temperature in dry air but reacts with water and readily corrodes; it also burns in the air when heated. Trivalent ions are used in the same way as some other rare earths in certain laser and glass-colorant applications. This is the case because holmium yields its own set of unique emission light lines from its fluorescent properties. Significantly, holmium has the highest magnetic permeability of any element and is therefore used for the polepieces of the strongest static magnets (also called a magnetic flux concentrator). Moreover, because holmium strongly absorbs nuclear fission-bred neutrons, it is also used as a burnable poison in nuclear reactors (Emsley 2001; Hoard et al. 1985). Holmium is also used as a colorant for glass, providing red or yellow hues. Holmium has no biological role in humans, but holmium salts are able to stimulate metabolism (Coldeway, 2017). The jury is still out on the long-term biological effects of holmium.

Personal Safety Precautions and Handling: Holmium[7]

HMIS Ratings: Health: 0 Flammability: 1 Physical: 1
NFPA Ratings: Health: 1 Flammability: 1 Reactivity: 1

Currently, holmium's OSHA PEL has not been established. Established engineering controls related to exposure and personal protection state that holmium should be within a humidity-controlled atmosphere and in an enclosed, controlled process under dry argon where possible. Ensure adequate ventilation to reduce exposure levels. Do not use food or drink in the work area. Wash thoroughly before eating or smoking. Do not blow dust off clothing or skin with compressed air. Because the metal dust presents a fire and explosion hazard prepare for the possibility of a fire. Keep extinguishing agents, tools for handling, and protective clothing readily available. Use only Class D dry powder extinguishing agent.

Recommended Handling Procedures for Holmium

Storage

Holmium does not oxidize at room temperature in air. However, for long-term storage, closed containers are recommended. Oils should not be used.

Cleaning

If surface oxidation has occurred due to exposure to acid fumes or slightly elevated temperatures, the major portion should be removed by filing and the final polishing electrolytically (see below).

Electropolishing

For Ho an electrolyte of 1% (or up to 6%) perchloric acid in absolute methanol is stirred and cooled continuously in a dry ice-acetone bath. A platinum cylinder (cup) serves as the cathode. A current density of about 0.5 amps/cm^2 usually is required. A variable voltage supply should be used and the amperage controlled to give small bubbles at the surface of the sample. The electrolyte should not be allowed to bubble excessively. The sample should be rinsed while cold in the dry-ice acetone bath, then rinsed with copious quantities of methanol.

Cutting

A metal saw (hack saw or jeweler's saw), or a low-speed diamond saw, or a spark cutter may be used. The metal should be electroplated after cutting since the freshly cut surface is quite reactive. Shearing is not recommended unless the sheared surface is filed off. The low-speed diamond saw or the spark cutter is recommended as the best method for obtaining a strain-free surface.

Cold Working

Ho can be cold swaged (i.e., spliced) or rolled about 30% reduction in cross-section without heat treatment. To prevent contamination, it should be wrapped or (even better) sealed in tantalum.

Handling

Because holmium reacts primarily with moisture, it should not be touched with bare hands, especially if they are to be heated. Plastic gloves are recommended. It can be handled in air. However, as mentioned above, strained surfaces from cutting or filing should be removed by electropolishing.

Stress Relief

The surface should be freshly cleaned by electropolishing just prior to heat treatment. A vacuum of 10 torr or better is required to prevent contamination. Minimal contamination will occur at 10 torr if the samples are wrapped in tantalum. The recommended temperature is half of the melting point in K for about 8 hours.

Melting

Ho may be arc or electron beam melted. Induction heating in outgassed tantalum or tungsten crucibles is most suitable. If holmium is heated in tantalum or tungsten to temperatures significantly above its melting point, tantalum and tungsten will dissolve in the molten rare earth.

ERBIUM

DETAILS

Figure 5.27 illustrates dysprosium's chemical symbol, element name, atomic number, and atomic weight; Figure 5.28 illustrates dysprosium's electron configuration.

Figures 5.29 and 5.30 show photographed samples of erbium.

Melting point—1802 K (1529°C, 2784°F)
Phase at standard temperature and pressure (STP)—solid
Appearance—silvery white
Boiling point—2868°C
Density—9.066 g/cm^3
Solubility in water—insoluble
Heat of fusion—19.90 kJ/mol
Vickers hardness—430–700 MPa

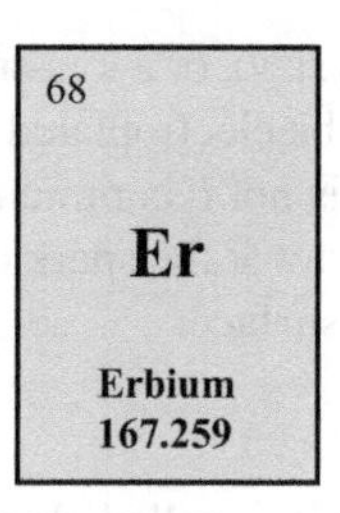

FIGURE 5.27 Chemical element designation for erbium with chemical symbol, element name, atomic number, and atomic weight.

$$\mathbf{Xe\ 4f^{12}\ 6s^{2}}$$

FIGURE 5.28 Erbium electron shell configuration.

FIGURE 5.29 Erbium.

FIGURE 5.30 Erbium.

Heat of vaporization—280 kJ/mol
Electronegativity—1.24
Natural occurrence—primordial
Magnetic ordering—paramagnetic at 300 K
Naming—originally found in a mine in Ytterby, Sweden, the source of the element's name
Curie point—32 K

Erbium, Er, is a silvery-white solid yet soft metal when isolated and is characterized by its pink-colored ions. Because of its pink color Erbium is sometimes used as a colorant for porcelain, glass (used in sunglasses and cheap jewelry), and cubic zirconia. It is always found in combination with other elements (Hammond, 2004; Stwertka, 1996). Erbium's principal use is based on its optical fluorescent properties which are especially useful in laser operations. Erbium is also used as an optical amplification media which is a simple laser amplifier of signals transmitted by fiber optics. Erbium is also used in a wide variety of medical applications including in dentistry and dermatology. Erbium is used in neutron-absorbing control rods in nuclear technology (Emsley, 2002). Erbium does not play a known biological role but is thought to stimulate the metabolism (Emsley, 2002).

Personal Safety Precautions and Handling: Holmium[8]

HMIS Ratings: Health: 0 Flammability: 1 Physical: 1
NFPA Ratings: Health: 1 Flammability: 1 Reactivity: 1

Currently, erbium's OSHA PEL has not been established. Established engineering controls related to exposure and personal protection state that erbium should be within a humidity-controlled atmosphere and in an enclosed, controlled process under dry argon where possible. Ensure adequate ventilation to reduce exposure levels. Do not use food or drink in work area. Wash thoroughly before eating or smoking. Do not blow dust off clothing or skin with compressed air. Because the metal dust presents a fire and explosion hazard prepare for the possibility of a fire. Keep extinguishing agents, tools for handling, and protective clothing readily available. Use only Class D dry powder extinguishing agent.

Recommended Handling Procedures for Erbium

Storage

Erbium does not oxidize at room temperature in air. However, for long-term storage, closed containers are recommended. Oils should not be used.

Cleaning

If surface oxidation has occurred due to exposure to acid fumes or slightly elevated temperatures, the major portion should be removed by filing and the final polishing electrolytically (see below).

Electropolishing

For Er, an electrolyte of 1% (or up to 6%) perchloric acid in absolute methanol is stirred and cooled continuously in a dry ice-acetone bath. A platinum cylinder (cup) serves as the cathode. A current density of about 0.5 amps/cm^2 usually is required. A variable voltage supply should be used and the amperage controlled to give small bubbles at the surface of the sample. The electrolyte should not be allowed to bubble excessively. The sample should be rinsed while cold in the dry-ice acetone bath, then rinsed with copious quantities of methanol.

Cutting

A metal saw (hack saw or jeweler's saw), or a low-speed diamond saw, or a spark cutter may be used. The metal should be electroplated after cutting since the freshly cut surface is quite reactive. Shearing is not recommended unless the sheared surface is filed off. The low-speed diamond saw or the spark cutter is recommended as the best method for obtaining a strain-free surface.

Cold Working

Er can be cold swaged (i.e., spliced) or rolled about 30% reduction in cross-section without heat treatment. To prevent contamination, it should be wrapped or (even better) sealed in tantalum.

Handling

Because erbium reacts primarily with moisture, it should not be touched with bare hands, especially if they are to be heated. Plastic gloves are recommended. It can be handled in air. However, as mentioned above, strained surfaces from cutting or filing should be removed by electropolishing.

Stress Relief

The surface should be freshly cleaned by electropolishing just prior to heat treatment. A vacuum of 10 torr or better is required to prevent contamination. Minimal contamination will occur at 10 torr if the samples are wrapped in tantalum. The recommended temperature is half of the melting point in K for about 8 hours.

Melting

Er may be arc or electron beam melted. Induction heating in outgassed tantalum or tungsten crucibles is most suitable. If erbium is heated in tantalum or tungsten to temperatures significantly above its melting point, tantalum and tungsten will dissolve in the molten rare earth.

THULIUM

Details

Figure 5.31 illustrates dysprosium's chemical symbol, element name, atomic number, and atomic weight; Figure 5.32 illustrates dysprosium's electron configuration.

Figures 5.33 and 5.34 show thulium samples.

Melting point—1818 K (1545°C, 2813°F)
Phase at standard temperature and pressure (STP)—solid
Appearance—silvery gray
Boiling point—2223°C
Density—9.32 g/cm^3
Solubility in water—insoluble
Heat of fusion—16.84 kJ/mol
Vickers hardness—470–650 MPa

FIGURE 5.31 Chemical element designation for erbium with chemical symbol, element name, atomic number, and atomic weight.

Xe 4f^{13}6s^{2}

FIGURE 5.32 Thulium electron shell configuration.

FIGURE 5.33 Thulium.

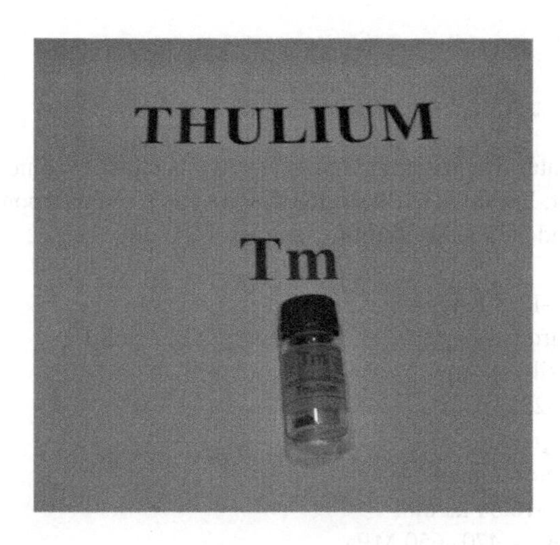

FIGURE 5.34 Thulium element. Photo by F. Spellman.

Heat of vaporization—191 kJ/mol
Electronegativity—1.25
Natural occurrence—primordial
Magnetic ordering—paramagnetic at 300 K
Naming—after Thule, a mythical region in Scandinavia
Curie point—25 K

Thulium, Tm, the second rarest or least abundant of the lanthanides—it is never found in nature in pure—and is a bright silvery-gray easily worked soft metal—it can be easily cut with a knife—and slowly tarnishes when isolated in air and is characterized by its pink-colored ions. Thulium is used as source in solid-state lasers and X-ray devices. Through ion exchange, thulium is principally extracted from monazite ores found in river sands. When soluble, thulium salts are somewhat toxic (Emsley, 2002). Thulium dust can cause explosions.

Personal Safety Precautions and Handling: Thulium[9]

HMIS Ratings: Health: 0 Flammability: 1 Physical: 1
NFPA Ratings: Health: 1 Flammability: 1 Reactivity: 1

Currently, thulium's OSHA PEL has not been established. Established engineering controls related to exposure and personal protection state that thulium should be a humidity-controlled atmosphere and in an enclosed, controlled process under dry argon where possible. Ensure adequate ventilation to reduce exposure levels. Do not use food or drink in work area. Wash thoroughly before eating or smoking. Do not blow dust off clothing or skin with compressed air. Because the metal dust presents a fire and explosion hazard prepare for the possibility of a fire. Keep extinguishing agents, tools for handling, and protective clothing readily available. Use only Class D dry powder extinguishing agent.

Recommended Handling Procedures for Thulium

Storage

Thulium does not oxidize at room temperature in air. However, for long-term storage, closed containers are recommended. Oils should not be used.

Cleaning

If surface oxidation has occurred due to exposure to acid fumes or slightly elevated temperatures, the major portion should be removed by filing and the final polishing electrolytically (see below).

Electropolishing

For Tm, an electrolyte of 1% (or up to 6%) perchloric acid in absolute methanol is stirred and cooled continuously in a dry ice-acetone bath. A platinum cylinder (cup) serves as the cathode. A current density of about 0.5 amps/cm^2 usually is required. A variable voltage supply should be used and the amperage controlled to give small bubbles at the surface of the sample. The electrolyte should not be allowed to bubble excessively. The sample should be rinsed while cold in the dry-ice acetone bath, then rinsed with copious quantities of methanol.

Cutting

A metal saw (hack saw or jeweler's saw), or a low-speed diamond saw, or a spark cutter may be used. The metal should be electroplated after cutting since the freshly cut surface is quite reactive. Shearing is not recommended unless the sheared surface is filed off. The low-speed diamond saw or the spark cutter is recommended as the best method for obtaining a strain-free surface.

Cold Working

Tm can be cold swaged (i.e., spliced) or rolled about 30% reduction in cross-section without heat treatment. To prevent contamination, it should be wrapped or (even better) sealed in tantalum.

Handling

Because thulium reacts primarily with moisture, it should not be touched with bare hands, especially if they are to be heated. Plastic gloves are recommended. It can be handled in air. However, as mentioned above, strained surfaces from cutting or filing should be removed by electropolishing.

Stress Relief

The surface should be freshly cleaned by electropolishing just prior to heat treatment. A vacuum of 10 torr or better is required to prevent contamination. Minimal contamination will occur at 10 torr if the samples are wrapped in tantalum. The recommended temperature is half of the melting point in K for about 8 hours.

Melting

Tm, due to its high pressure, can't be melted and if arc welded will lose great amounts of metal. Induction heating in outgassed tantalum or tungsten crucibles is most suitable. If thulium is heated in tantalum or tungsten to temperatures significantly above its melting point, tantalum and tungsten will dissolve in the molten rare earth.

YTTERBIUM

Details

Figure 5.35 illustrates ytterbium's chemical symbol, element name, atomic number, and atomic weight; Figure 5.36 illustrates ytterbium's electron configuration.

Figures 5.37 and 5.38 show the element ytterbium.

Melting point—1097 K (824°C, 1515°F)
Phase at standard temperature and pressure (STP)—solid
Appearance—silvery white with pale yellow tint
Boiling point—1196°C
Density—6.90 g/cm^3

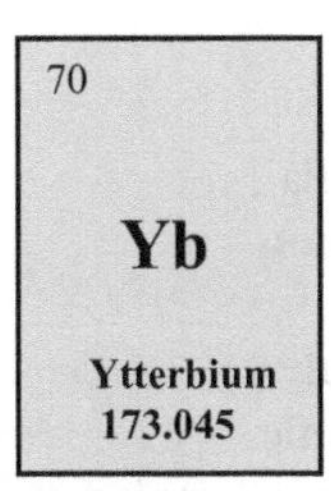

FIGURE 5.35 Chemical element designation for ytterbium with chemical symbol, element name, atomic number, and atomic weight.

Xe 4f^{14} 6s^{2}

FIGURE 5.36 Ytterbium electron shell configuration.

FIGURE 5.37 Ytterbium.

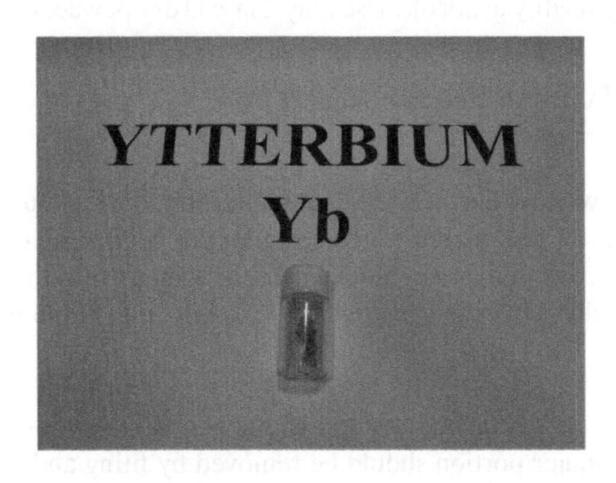

FIGURE 5.38 Photo of element ytterbium. Photo by F. Spellman.

Solubility in water—insoluble
Heat of fusion—7.66 kJ/mol
Vickers hardness—205–250 MPa
Heat of vaporization—129 kJ/mol
Electronegativity—1.1
Natural occurrence—primordial
Magnetic ordering—paramagnetic
Naming—after Ytterby, Sweden, where it was mined
Curie point—NA

Ytterbium, Yb, the 14th and penultimate member of the lanthanide series is significantly different from most of the other lanthanides because of its closed-shell electron configuration. It is a soft, malleable, ductile chemical element that displays, in pure form, a bright silvery luster. It oxidizes slowly in air and reacts slowly in cold water (Hammond, 2005). Ytterbium is found with other rare-earth elements in several rare mins and is most often recovered from monazite sand. It is commonly used in infrared lasers, rechargeable batteries, fiber optics, and as a chemical reducing agent.

Personal Safety Precautions and Handling: Ytterbium[10]

HMIS Ratings: Health: 0 Flammability: 1 Physical: 1
NFPA Ratings: Health: 1 Flammability: 1 Reactivity: 1

Currently, ytterbium's OSHA PEL has not been established. Established engineering controls related to exposure and personal protection state that ytterbium should be a humidity-controlled atmosphere and in an enclosed, controlled process under dry argon where possible. Ensure adequate ventilation to reduce exposure levels. Do not use food or drink in work area. Wash thoroughly before eating or smoking. Do not blow dust off clothing or skin with compressed air. Once extracted and processed, ytterbium is somewhat hazardous as an eye and skin irritant. Moreover, the metal is a fire and explosion hazard. Because the metal dust presents a fire and explosion hazard prepare for the possibility of a fire. Keep extinguishing agents, tools for handling, and protective clothing readily available. Use only Class D dry powder extinguishing agent.

Recommended Handling Procedures for Yb

Storage

Ytterbium will slowly oxidize at room temperature in air. It should be stored under 10 torr [i.e., 1 torr = 1 mm mercury (0°C) mmHg] or better vacuum or in sealed jar under an inert gas. For long-term storage, the best method is to seal this metal in evacuated Pyrex tubes with the ends fused by fusion. Oils should not be used.

Cleaning

If surface oxidation has occurred due to exposure to acid fumes or slightly elevated temperatures, the major portion should be removed by filing and the final polishing electrolytically (see below).

Electropolishing

For Yb, a chemical polis of 5–8 mL HNO, 58 mL HPO, and 22 mL of ethanol swabbed on for 10 seconds works well. Electropolishing with 10 vol.% HCL in methanol at room temperature also works.

Cutting

A metal saw (hack saw or jeweler's saw), or a low-speed diamond saw, or a spark cutter may be used. The metal should be electroplated after cutting since the freshly cut surface is quite reactive. Shearing is not recommended unless the sheared surface is filed off. The low-speed diamond saw or the spark cutter is recommended as the best method for obtaining a strain-free surface.

Cold Working

Yb can be cold worked 50% or more without heat treatment. To prevent contamination, it should be wrapped or (even better) sealed in tantalum.

Handling

Because ytterbium reacts primarily with moisture, it should not be touched with bare hands, especially if they are to be heated. Plastic gloves are recommended. It can be handled in air, but an oxide layer does form quite quickly. This layer can be removed and the surface passivated by electropolishing.

Stress Relief

The surface should be freshly cleaned by electropolishing just prior to heat treatment. A vacuum of 10 torr or better is required to prevent contamination. Minimal contamination will occur at 10 torr if the samples are wrapped in tantalum. The recommended temperature is half of the melting point in K for about 8 hours.

Melting

Yb should not be melted.

LUTETIUM

Details

Figure 5.39 illustrates lutetium's chemical symbol, element name, atomic number, and atomic weight; Figure 5.40 illustrates lutetium's electron configuration.

Figures 5.41 and 5.42 show samples of lutetium.

Melting point—1925 K (1652°C, 3006°F)
Phase at standard temperature and pressure (STP)—solid
Appearance—silvery white
Boiling point—3402°C
Density—9.841 g/cm^3
Solubility in water—insoluble
Heat of fusion—ca 22 kJ/mol

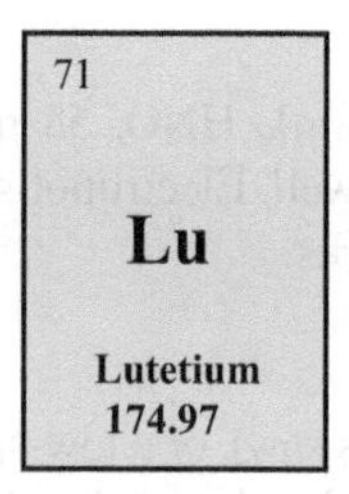

FIGURE 5.39 Chemical element designation for lutetium with chemical symbol, element name, atomic number, and atomic weight.

$$\mathbf{Xe\ 4f^{14}\ 5d^{1}\ 6s^{2}}$$

FIGURE 5.40 Lutetium electron shell configuration.

FIGURE 5.41 Lutetium.

FIGURE 5.42 Lutetium. Photo by F. Spellman.

Vickers hardness—755–1160 MPa
Heat of vaporization—414 kJ/mol
Electronegativity—1.27
Natural occurrence—primordial
Magnetic ordering—paramagnetic
Naming—after Lutetia, Latin for Paris, in the Roman era
Curie point—NA

Lutetium, Lu, the last element in the lanthanide series, is a silvery white metal, which resists corrosion in dry air, but not in moist air. Lutetium is not an abundant element but is appreciably more common than silver in the earth's crust. Lutetium usually occurs linked with the element yttrium and is sometimes used in PET scan detectors, superconductors, high refractive index glass, and X-ray phosphor, although it is not the element of the first choice due to its high price. Lutetium is regarded as having a low degree of toxicity, but caution is advised when handling because inhalation is dangerous and its compounds can cause skin irritation (Krebs, 2006).

Personal Safety Precautions and Handling: Lutetium[11]

HMIS Ratings: Health: 0 Flammability: 1 Physical: 1
NFPA Ratings: Health: 1 Flammability: 1 Reactivity: 1

Currently, lutetium's OSHA PEL has not been established. Established engineering controls related to exposure and personal protection state that lutetium should be a humidity-controlled atmosphere and in an enclosed, controlled process under dry argon where possible. Ensure adequate ventilation to reduce exposure levels. Do not use food or drink in work area. Wash thoroughly before eating or smoking. Do not blow dust off clothing or skin with compressed air. Once extracted and processed, lutetium is somewhat hazardous as an eye and skin irritant. Moreover, the metal is a fire and explosion hazard. Because the metal dust presents a fire and explosion hazard prepare for the possibility of a fire. Keep extinguishing agents, tools for handling, and protective clothing readily available. Use only Class D dry powder extinguishing agent.

Recommended Handling Procedures for Lu

Storage

Lutetium does not oxidize at room temperature in air. However, for long-term storage, closed containers are recommended. Oils should not be used.

Cleaning

If surface oxidation has occurred due to exposure to acid fumes or slightly elevated temperatures, the major portion should be removed by filing and the final polishing electrolytically (see below).

Electropolishing

For Lu, an electrolyte of 1% (or up to 6%) perchloric acid in absolute methanol is stirred and cooled continuously in a dry ice-acetone bath. A platinum cylinder (cup) serves as the cathode. A current density of about 0.5 amps/cm^2 usually is required. A variable voltage supply should be used and the amperage controlled to give small bubbles at the surface of the sample. The electrolyte should not be allowed to bubble excessively. The sample should be rinsed while cold in the dry-ice acetone bath, then rinsed with copious quantities of methanol.

Cutting

A metal saw (hack saw or jeweler's saw), or a low-speed diamond saw, or a spark cutter may be used. The metal should be electroplated after cutting since the freshly cut surface is quite reactive. Shearing is not recommended unless the sheared surface is filed off. The low-speed diamond saw or the spark cutter is recommended as the best method for obtaining a strain-free surface.

Cold Working

Lu can be cold swaged (i.e., spliced) or rolled about 30% reduction in cross-section without heat treatment. To prevent contamination, it should be wrapped or (even better) sealed in tantalum.

Handling

Because lutetium reacts primarily with moisture, it should not be touched with bare hands, especially if they are to be heated. Plastic gloves are recommended. It can be handled in air. However, as mentioned above, strained surfaces from cutting or filing should be removed by electropolishing.

Stress Relief

The surface should be freshly cleaned by electropolishing just prior to heat treatment. A vacuum of 10 torr or better is required to prevent contamination. Minimal contamination will occur at 10 torr if the samples are wrapped in tantalum. The recommended temperature is half of the melting point in K for about 8 hours.

Melting

Lu may be arc or electron beam melted. Induction heating in outgassed tantalum or tungsten crucibles is most suitable. If lutetium is heated in tantalum or tungsten to temperatures significantly above its melting point, tantalum and tungsten will dissolve in the molten rare earth.

CERIUM

Details

Figure 5.43 illustrates cerium's chemical symbol, element name, atomic number, and atomic weight; Figure 5.44 illustrates cerium's electron configuration.

Figures 5.45 and 5.46 show samples of cerium.

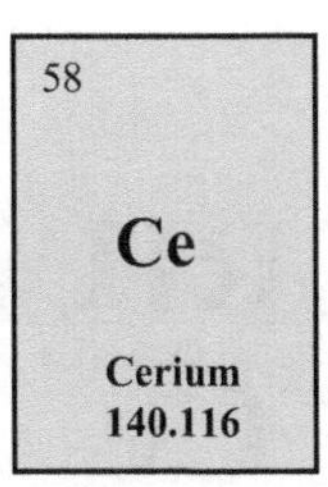

FIGURE 5.43 Chemical element designation for cerium with chemical symbol, element name, atomic number, and atomic weight.

Xe $4f^1$ $5d^1$ $6s^2$

FIGURE 5.44 Cerium electron shell configuration.

FIGURE 5.45 Cerium.

Melting point—1068 K (795°C, 1463°F)
Phase at standard temperature and pressure (STP)—solid
Appearance—silvery white
Boiling point—3443°C
Density—6.770 g/cm^3
Solubility in water—decomposes
Heat of fusion—5.46 kJ/mol
Vickers hardness—210–470 MPa
Heat of vaporization—398 kJ/mol
Electronegativity—1.12
Natural occurrence—primordial
Magnetic ordering—paramagnetic
Naming—after dwarf planet Ceres, Roman deity of Ceres
Curie point—NA

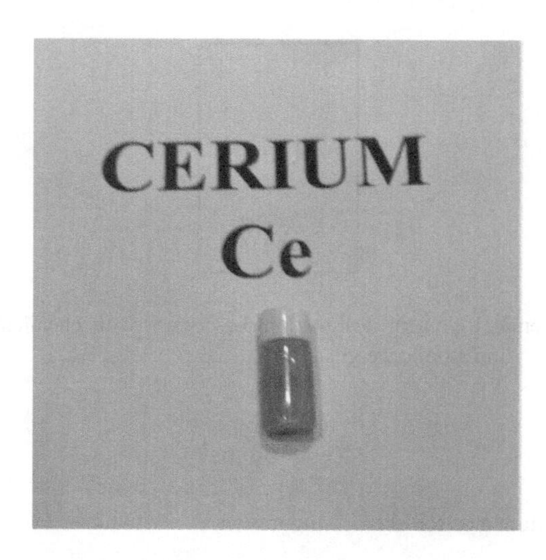

FIGURE 5.46 Photo of cerium element stored in liquid to prevent exposure to air.

Photo by F. Spellman.

Cerium, Ce, is the second element in the lanthanoid series and an f-block element whose turnings or gritty powder that appears as a gray metallic solid that tarnishes when exposed to air, forming a spalling layer (meaning it flakes off in fragments) similar to iron rust. It is soft and can be cut with a steel kitchen knife and is the most common of the lanthanoids. Cerium has no biological role as humans and is not very toxic. Compared to other rare-earths cerium is easier to extract from its ores because of it ability to be oxidized to the +4 state; it is also five times more abundant in earth's crust than lead and more common than tin. Cerium and its compounds have a variety of uses: for example, it is used as a catalyst, metal alloy, water purifier, radiation shield, cerium oxide is used to polish glass, is a key ingredient in catalytic converters and cerium metal is used in lighters for its pyrophoric qualities.

Personal Safety Precautions and Handling: Cerium[12]

HMIS Ratings: Health: 1 Flammability: 3 Physical: 1
NFPA Ratings: Health: 1 Flammability: 3 Reactivity: 1

Currently, cerium's OSHA PEL has not been established. Established engineering controls related to exposure and personal protection state that cerium should be a humidity-controlled atmosphere and in an enclosed, controlled process under dry argon where possible. Ensure adequate ventilation to reduce exposure levels. Do not use food or drink in work area. Contact may burn skin, eyes, or mucous. Wash thoroughly before eating or smoking. Do not blow dust off clothing or skin with compressed air. Once extracted and processed, cerium is somewhat hazardous as an eye and skin irritant. Moreover, the metal is a fire and explosion hazard. Because the metal dust presents a fire and explosion hazard prepare for the possibility of a fire.

Keep extinguishing agents, tools for handling, and protective clothing readily available. Use only Class D dry powder extinguishing agent.

Recommended Handling Procedures for Ce

Storage

Cerium will slowly oxidize at room temperature in air. It should be stored under 10 torr [i.e., 1 torr = 1 mm mercury (0°C) mmHg] or better vacuum or in sealed jar under vacuum. For long-term storage, the best method is to seal this metal in evacuated Pyrex tubes with the ends fused by fusion. Oils should not be used.

Cleaning

Even when stored as described above, some surface oxidation will occur. The oxidation should be removed by filing. A wire-brush wheel may be used, but filing is preferred. If the surface has almost turned white, at least one mm of metal should be removed to ensure the removal of intergranular corrosion products which are near the surface. After filing, the cold worked surface can be removed by electropolishing (see below) which also passivates the surface.

Electropolishing

An electrolyte of 1% (or up to 6%) perchloric acid in absolute methanol is stirred and cooled continuously in a dry ice-acetone bath. A platinum cylinder (cup) serves as the cathode. A current density of about 0.5 amps/cm^2 usually is required. A variable voltage supply should be used and the amperage controlled to give small bubbles at the surface of the sample. The electrolyte should not be allowed to bubble excessively. The sample should be rinsed while cold in the dry-ice acetone bath, then rinsed with copious quantities of methanol.

Cutting

A metal saw (hack saw or jeweler's saw), or a low-speed diamond saw, or a spark cutter may be used. The metal should be electroplated after cutting since the freshly cut surface is quite reactive. Shearing is not recommended unless the sheared surface is filed off. The low-speed diamond saw or the spark cutter is recommended as the best method for obtaining a strain-free surface.

Cold Working

Cerium can be cold swaged (i.e., spliced) or rolled about 30% reduction in cross-section without heat treatment. To prevent contamination, it should be wrapped or (even better) sealed in tantalum.

Handling

Because cerium reacts primarily with moisture, it should not be touched with bare hands. Plastic gloves are recommended. It can be handled in air, but an oxide layer does form quite quickly. This layer can be removed and the surface passivated by electropolishing.

Stress Relief

The surface should be freshly cleaned by electropolishing just prior to heat treatment. A vacuum of 10 torr or better is required to prevent contamination. Minimal contamination will occur at 10 torr if the samples are wrapped in tantalum. The recommended temperature is half of the melting point in K for about 8 hours.

Melting

Cerium may be arc or electron beam melted. Levitation and induction hearting in outgassed tantalum or tungsten crucibles are also suitable. If cerium is heated in tantalum or tungsten to temperatures significantly above their melting points, tantalum and tungsten will dissolve in the molten rare earth.

PRASEODYMIUM

Details

Figure 5.47 illustrates cerium's chemical symbol, element name, atomic number, and atomic weight; Figure 5.48 illustrates cerium's electron configuration.

Figures 5.49 and 5.50 show samples of praseodymium.

Melting point—1208 K (3130°C, 5666°F)
Phase at standard temperature and pressure (STP)—solid
Appearance—grayish white
Boiling point—3130°C
Density—6.77 g/cm^3
Solubility in water—decomposes
Heat of fusion—6.89 kJ/mol
Vickers hardness—210–470 MPa
Heat of vaporization—331 kJ/mol

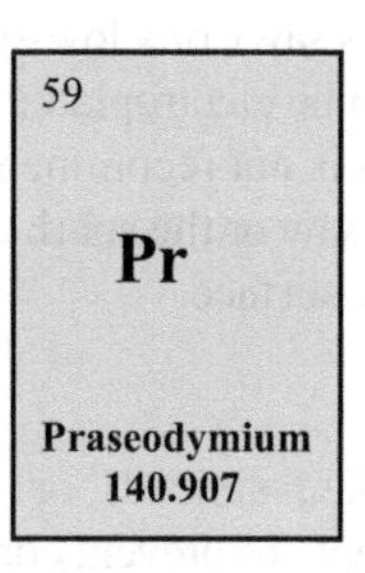

FIGURE 5.47 Chemical element designation for praseodymium with chemical symbol, element name, atomic number, and atomic weight.

Xe 4f^3 6s^2

FIGURE 5.48 Illustrates praseodymium's electron configuration.

FIGURE 5.49 Praseodymium.

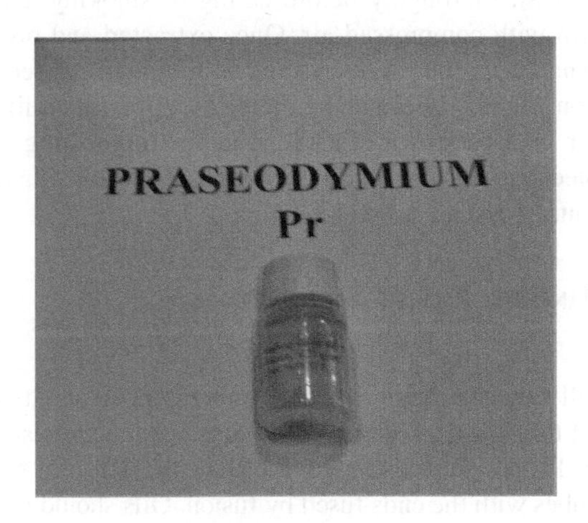

FIGURE 5.50 Photo of element praseodymium in liquid to prevent tarnishing.

Electronegativity—1.13
Natural occurrence—primordial
Magnetic ordering—paramagnetic
Naming—from the Greek meaning "green twin"
Curie point—NA

Praseodymium, Pr, is a silvery, soft, malleable, and ductile metal (with hardness comparable silver), prized for magnetic, electrical, chemical, and optical properties, and many of its industrial uses involve its ability to filter yellow light from light sources. Praseodymium is too reactive to be found in natural form; when in pure form and exposed in air, it develops a green oxide coating. Praseodymium applications include use as magnets, lasers, pigments, and cryogenic refrigerants. More specifically, praseodymium in magnet use finds applications in small equipment, such as printers, watches, motors, headphones, loudspeakers, and magnetic storage (McGill, 2021). Praseodymium compounds give enamels, glasses, and ceramics a yellow color (Lide, 2007; McGill, 2021).

Personal Safety Precautions and Handling[13]

HMIS Ratings: Health: 1 Flammability: 1 Physical: 1
NFPA Ratings: Health: 1 Flammability: 1 Reactivity: 1

Currently, praseodymium's OSHA PEL has not been established. Established engineering controls related to exposure and personal protection state that praseodymium should be a humidity-controlled atmosphere and in an enclosed, controlled process under dry argon where possible. Ensure adequate ventilation to reduce exposure levels. Do not use food or drink in work area. Contact may burn skin, eyes, or mucous. Wash thoroughly before eating or smoking. Do not blow dust off clothing or skin with compressed air. Once extracted and processed, praseodymium is somewhat hazardous as an eye and skin irritant. Moreover, the metal is a fire and explosion hazard. Because the metal dust presents a fire and explosion hazard prepare for the possibility of a fire. Keep extinguishing agents, tools for handling and protective clothing readily available. Use only Class D dry powder extinguishing agent.

Recommended Handling Procedures for Pr

Storage

Praseodymium will slowly oxidize at room temperature in air. It should be stored under 10 torr [i.e., 1 torr = 1 mm mercury (0°C) mmHg] or better vacuum or in sealed jar under vacuum. For long-term storage, the best method is to seal this metal in evacuated Pyrex tubes with the ends fused by fusion. Oils should not be used.

Cleaning

Even when stored as described above, some surface oxidation will occur. The oxidation should be removed by filing. A wire-brush wheel may be used, but filing is preferred. If the surface has almost turned white, at least one mm of metal should be removed to ensure the removal of intergranular corrosion products which are near the surface. After filing, the cold worked surface can be removed by electropolishing (see below) which also passivates the surface.

Electropolishing

An electrolyte of 1% (or up to 6%) perchloric acid in absolute methanol is stirred and cooled continuously in a dry ice-acetone bath. A platinum cylinder (cup) serves as the cathode. A current density of about 0.5 amps/cm^2 usually is required. A variable voltage supply should be used and the amperage controlled to give small bubbles at the surface of the sample. The electrolyte should not be allowed to bubble excessively. The sample should be rinsed while cold in the dry-ice acetone bath, then rinsed with copious quantities of methanol.

Cutting

A metal saw (hack saw or jeweler's saw), or a low-speed diamond saw, or a spark cutter may be used. The metal should be electroplated after cutting since the freshly cut surface is quite reactive. Shearing is not recommended unless the sheared surface is filed off. The low-speed diamond saw or the spark cutter is recommended as the best method for obtaining a strain-free surface.

Cold Working

Praseodymium can be cold swaged (i.e., spliced) or rolled about 30% reduction in cross-section without heat treatment. To prevent contamination, it should be wrapped or (even better) sealed in tantalum.

Handling

Because praseodymium reacts primarily with moisture, it should not be touched with bare hands. Plastic gloves are recommended. It can be handled in air, but an oxide layer does form quite quickly. This layer can be removed and the surface passivated by electropolishing.

Stress Relief

The surface should be freshly cleaned by electropolishing just prior to heat treatment. A vacuum of 10 torr or better is required to prevent contamination. Minimal contamination will occur at 10 torr if the samples are wrapped in tantalum. The recommended temperature is half of the melting point in K for about 8 hours.

Melting

Praseodymium may be arc or electron beam melted. Levitation and induction hearting in outgassed tantalum or tungsten crucibles are also suitable. If praseodymium is heated in tantalum or tungsten to temperatures significantly above their melting points, tantalum and tungsten will dissolve in the molten rare earth.

LANTHANUM

Details

Figure 5.51 illustrates cerium's chemical symbol, element name, atomic number, and atomic weight; Figure 5.52 illustrates cerium's electron configuration.

Figures 5.53 and 5.54 show the element lanthanum.

Melting point—1193 K (920°C, 1688°F)
Phase at standard temperature and pressure (STP)—solid

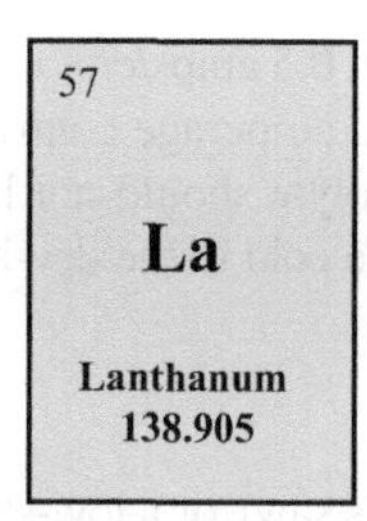

FIGURE 5.51 Chemical element designation for Lanthanum with chemical symbol, element name, atomic number, and atomic weight.

Xe $5d^1$ $6s^2$

FIGURE 5.52 Illustrates lanthanum's electron configuration.

FIGURE 5.53 Lanthanum.

FIGURE 5.54 Photo of the rare-earth element lanthanum. Photo by F. Spellman.

Appearance—silvery white
Boiling point—3464°C
Density—6.162 g/cm³
Solubility in water—decomposes
Heat of fusion—6.20 kJ/mol
Vickers hardness—360–1750 MPa
Heat of vaporization—400 kJ/mol
Electronegativity—1.10
Natural occurrence—primordial
Magnetic ordering—weakly paramagnetic
Naming—from the Greek meaning "to lie hidden"
Curie point—NA

Lanthanum, La, is a silvery-white, soft, malleable, and ductile metal that tarnishes slowly when exposed to air. Lanthanum is valued for applications as petroleum refining catalysts, additives in optical glass, carbon arc lamps for projectors and studio lights, batteries, camera lenses, ignition elements in torches and lighters, scintillators, welding electrodes, and several other things. Lanthanum has no know biological role in humans.

Personal Safety Precautions and Handling[14]

HMIS Ratings: Health: 1 Flammability: 2 Physical: 1
NFPA Ratings: Health: 1 Flammability: 2 Reactivity: 1

Currently, lanthanum's OSHA PEL has not been established. Established engineering controls related to exposure and personal protection state that lanthanum should be a humidity-controlled atmosphere and in an enclosed, controlled process under dry argon where possible. Ensure adequate ventilation to reduce exposure levels. Do not use food or drink in work area. Contact may burn skin, eyes, or mucous. Wash thoroughly before eating or smoking. Do not blow dust off clothing or skin with compressed air. Once extracted and processed, lanthanum is somewhat hazardous as an eye and skin irritant. Moreover, the metal is a fire and explosion hazard. Because the metal dust presents a fire and explosion hazard prepare for the possibility of a fire. Keep extinguishing agents, tools for handling, and protective clothing readily available. Use only Class D dry powder extinguishing agent—do not use water.

Recommended Handling Procedures for La

Storage

Lanthanum will slowly oxidize at room temperature in air. It should be stored under 10 torr [i.e., 1 torr = 1 mm mercury (0°C) mmHg] or better vacuum or in sealed jar under vacuum. For long-term storage, the best method is to seal this metal in evacuated Pyrex tubes with the ends fused by fusion. Oils should not be used.

Cleaning

Even when stored as described above, some surface oxidation will occur. The oxidation should be removed by filing. A wire-brush wheel may be used, but filing is preferred. If the surface has almost turned white, at least one mm of metal should be removed to ensure the removal of intergranular corrosion products which are near the surface. After filing, the cold worked surface can be removed by electropolishing (see below) which also passivates the surface.

Electropolishing

An electrolyte of 1% (or up to 6%) perchloric acid in absolute methanol is stirred and cooled continuously in a dry ice-acetone bath. A platinum cylinder (cup) serves as the cathode. A current density of about 0.5 amps/cm^2 usually is required. A variable voltage supply should be used and the amperage controlled to give small bubbles at the surface of the sample. The electrolyte should not be allowed to bubble excessively. The sample should be rinsed while cold in the dry-ice acetone bath, then rinsed with copious quantities of methanol.

Cutting

A metal saw (hack saw or jeweler's saw), or a low-speed diamond saw, or a spark cutter may be used. The metal should be electroplated after cutting since the freshly cut surface is quite reactive. Shearing is not recommended unless the sheared surface is filed off. The low-speed diamond saw or the spark cutter is recommended as the best method for obtaining a strain-free surface.

Cold Working

Lanthanum can be cold swaged (i.e., spliced) or rolled about 30% reduction in cross-section without heat treatment. To prevent contamination, it should be wrapped or (even better) sealed in tantalum.

Handling

Because lanthanum reacts primarily with moisture, it should not be touched with bare hands. Plastic gloves are recommended. It can be handled in air, but an oxide layer does form quite quickly. This layer can be removed and the surface passivated by electropolishing.

Stress Relief

The surface should be freshly cleaned by electropolishing just prior to heat treatment. A vacuum of 10 torr or better is required to prevent contamination. Minimal contamination will occur at 10 torr if the samples are wrapped in tantalum. The recommended temperature is half of the melting point in K for about 8 hours.

Melting

Lanthanum may be arc or electron beam melted. Levitation and induction hearting in outgassed tantalum or tungsten crucibles are also suitable. If lanthanum is heated in tantalum or tungsten to temperatures significantly above their melting points, tantalum and tungsten will dissolve in the molten rare earth.

YTTRIUM

Details

Figure 5.55 illustrates yttrium's chemical symbol, element name, atomic number, and atomic weight; Figure 5.56 illustrates yttrium's electron configuration.

Figures 5.57 and 5.58 show photos of the rare-earth element yttrium.

Melting point—1799 K (1526°C, 2779°F)
Phase at standard temperature and pressure (STP)—solid
Appearance—silvery white
Boiling point—2930°C
Density—4.472 g/cm^3
Solubility in water—Insoluble
Heat of fusion—11.42 kJ/mol
Brinell hardness—200–589 MPa

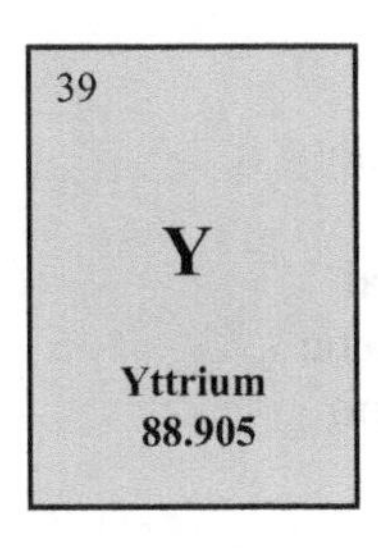

FIGURE 5.55 Chemical element designation for yttrium with chemical symbol, element name, atomic number, and atomic weight.

Kr 4d^1 5s^2

FIGURE 5.56 Illustrates yttrium's electron configuration.

FIGURE 5.57 Yttrium.

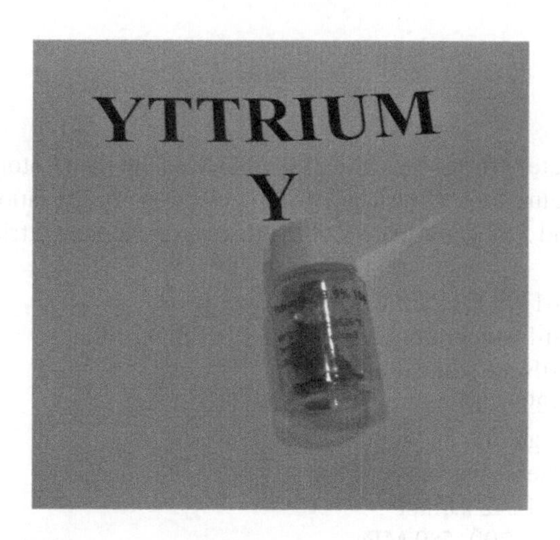

FIGURE 5.58 Yttrium. Photo by F. Spellman.

Heat of vaporization—363 kJ/mol
Electronegativity—1.22
Natural occurrence—primordial
Magnetic ordering—paramagnetic
Naming—after Ytterby (Sweden)
Curie point—NA

Yttrium, Y, is a silvery-metallic, soft, lustrous, and highly transition metal. It is relatively stable in air in pure, bulk form, but when finely divided yttrium is very unstable in air; its shavings and turnings can ignite in air at temperatures exceeding 400°C. Yttrium applications include ceramics, high-temperature superconductors, rechargeable batteries, TV phosphors that are also an important component of white LEDs and metal alloys. One of the major uses of yttrium in small quantities as the cathode is in lithium batteries where this type of battery offers high energy, long life, and safety. Yttrium has no known biological role, but it can be highly toxic to almost all lifeforms. Yttrium dust is highly flammable.

Personal Safety Precautions and Handling[15]

HMIS Ratings: Health: 0 Flammability: 1 Physical: 1
NFPA Ratings: Health: 1 Flammability: 1 Reactivity: 1

Currently, yttrium's OSHA PEL has not been established. Established engineering controls related to exposure and personal protection state that yttrium should be a humidity-controlled atmosphere and in an enclosed, controlled process under dry argon where possible. Ensure adequate ventilation to reduce exposure levels. Do not use food or drink in work area. Contact may burn skin, eyes, or mucous. Wash thoroughly before eating or smoking. Do not blow dust off clothing or skin with compressed air. Once extracted and processed, yttrium is somewhat hazardous as an eye and skin irritant. Moreover, the metal is a fire and explosion hazard. Because the metal dust presents a fire and explosion hazard prepare for the possibility of a fire. Keep extinguishing agents, tools for handling, and protective clothing readily available. Use only Class D dry powder extinguishing agent—do not use water.

Recommended Handling Procedures for Y

Storage

Yttrium does not oxidize at room temperature in air. However, for long-term storage, closed containers are recommended. Oils should not be used.

Cleaning

If surface oxidation has occurred due to exposure to acid fumes or slightly elevated temperatures, the major portion should be removed by filing and the final polishing electrolytically (see below).

Electropolishing

For Lu, an electrolyte of 1% (or up to 6%) perchloric acid in absolute methanol is stirred and cooled continuously in a dry ice-acetone bath. A platinum cylinder (cup) serves as the cathode. A current density of about 0.5 amps/cm^2 usually is required. A variable voltage supply should be used and the amperage controlled to give small bubbles at the surface of the sample. The electrolyte should not be allowed to bubble excessively. The sample should be rinsed while cold in the dry-ice acetone bath, then rinsed with copious quantities of methanol.

Cutting

A metal saw (hack saw or jeweler's saw), or a low-speed diamond saw, or a spark cutter may be used. The metal should be electroplated after cutting since the freshly cut surface is quite reactive. Shearing is not recommended unless the sheared surface is filed off. The low-speed diamond saw or the spark cutter is recommended as the best method for obtaining a strain-free surface.

Cold Working

Lu can be cold swaged (i.e., spliced) or rolled about 30% reduction in cross-section without heat treatment. To prevent contamination, it should be wrapped or (even better) sealed in tantalum.

Handling

Because lutetium reacts primarily with moisture, it should not be touched with bare hands, especially if they are to be heated. Plastic gloves are recommended. It can be handled in air. However, as mentioned above, strained surfaces from cutting or filing should be removed by electropolishing.

Stress Relief

The surface should be freshly cleaned by electropolishing just prior to heat treatment. A vacuum of 10 torr or better is required to prevent contamination. Minimal contamination will occur at 10 torr if the samples are wrapped in tantalum. The recommended temperature is half of the melting point in K for about 8 hours.

Melting

Lu may be arc or electron beam melted. Induction heating in outgassed tantalum or tungsten crucibles is most suitable. If lutetium is heated in tantalum or tungsten to temperatures significantly above its melting point, tantalum and tungsten will dissolve in the molten rare earth.

SCANDIUM

Details

Figure 5.59 illustrates scandium's chemical symbol, element name, atomic number, and atomic weight; Figure 5.60 illustrates scandium's electron configuration.

Figures 5.61 and 5.62 show photos of the rare-earth element scandium.

Melting point—1814 K (1541°C, 2806°F)
Phase at standard temperature and pressure (STP)—solid
Appearance—silvery white
Boiling point—2836°C
Density—2.985 g/cm^3
Solubility in water—Insoluble
Heat of fusion—14.1 kJ/mol
Brinell hardness—736–1200 MPa
Heat of vaporization—332.7 kJ/mol
Electronegativity—1.36
Natural occurrence—primordial
Magnetic ordering—paramagnetic
Naming—after Scandinavia
Curie point—NA

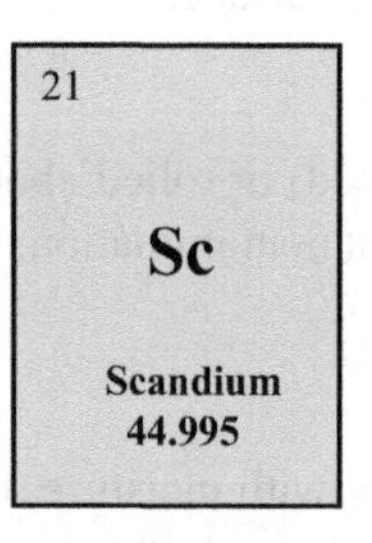

FIGURE 5.59 Chemical element designation for scandium with chemical symbol, element name, atomic number, and atomic weight.

Ar $3d^1 4s^2$

FIGURE 5.60 Illustrates scandium's electron configuration.

FIGURE 5.61 Scandium.

FIGURE 5.62 Photo of the rare-earth element scandium. Photo by F. Spellman.

Scandium, Sc, is a silvery-white metallic, soft, lustrous and develops a slightly pinkish or yellowish cast when oxidized in air. Scandium is found in deposits other rare earths and in uranium compounds. Initially, scandium was mined in only a few locations and was rather scarce with no known important application. However, by the 1970s when the confirmed effects of scandium on aluminum alloys were determined, and its use in such alloys, super alloys, became one of its major applications. Scandium is also used in semiconductors, baseball bats, lights, X-ray tubes, and in the manufacture of ultra-lite aerospace components. Scandium is prone to weathering and dissolves slowly in most dilute acids. Scandium turnings ignite in the air with a brilliant yellow flame to form scandium oxide (LANL, 2013).

Personal Safety Precautions and Handling[16]

HMIS Ratings: Health: 0 Flammability: 1 Physical: 1
NFPA Ratings: Health: 1 Flammability: 1 Reactivity: 1

Currently, scandium's OSHA PEL has not been established. Established engineering controls related to exposure and personal protection state that scandium should be a humidity-controlled atmosphere and in an enclosed, controlled process under dry argon where possible. Ensure adequate ventilation to reduce exposure levels. Do not use food or drink in work area. Contact may burn skin, eyes or mucous. Wash thoroughly before eating or smoking. Do not blow dust off clothing or skin with compressed air. Once extracted and processed, scandium is somewhat hazardous as an eye and skin irritant. Moreover, the metal is a fire and explosion hazard. Because the metal dust presents a fire and explosion hazard prepare for the possibility of a fire. Keep extinguishing agents, tools for handling and protective clothing readily available. Use only Class D dry powder extinguishing agent—do not use water.

Recommended Handling Procedures for Sc

Storage

Scandium does not oxidize at room temperature in air. However, for long-term storage, closed containers are recommended. Oils should not be used.

Cleaning

If surface oxidation has occurred due to exposure to acid fumes or slightly elevated temperatures, the major portion should be removed by filing and the final polishing electrolytically (see below).

Electropolishing

For Sc, an electrolyte of 1% (or up to 6%) perchloric acid in absolute methanol is stirred and cooled continuously in a dry ice-acetone bath. A platinum cylinder (cup) serves as the cathode. A current density of about 0.5 amps/cm^2 usually is required. A variable voltage supply should be used and the amperage controlled to give small bubbles at the surface of the sample. The electrolyte should not be allowed to bubble excessively. The sample should be rinsed while cold in the dry-ice acetone bath, then rinsed with copious quantities of methanol.

Cutting

A metal saw (hack saw or jeweler's saw), or a low-speed diamond saw, or a spark cutter may be used. The metal should be electroplated after cutting since the freshly cut surface is quite reactive. Shearing is not recommended unless the sheared surface is filed off. The low-speed diamond saw or the spark cutter is recommended as the best method for obtaining a strain-free surface.

Cold Working

Sc can be cold swaged (i.e., spliced) or rolled about 30% reduction in cross-section without heat treatment. To prevent contamination, it should be wrapped or (even better) sealed in tantalum.

Handling

Because scandium reacts primarily with moisture, it should not be touched with bare hands, especially if they are to be heated. Plastic gloves are recommended. It can be handled in air. However, as mentioned above, strained surfaces from cutting or filing should be removed by electropolishing.

Stress Relief

The surface should be freshly cleaned by electropolishing just prior to heat treatment. A vacuum of 10 torr or better is required to prevent contamination. Minimal contamination will occur at 10 torr if the samples are wrapped in tantalum. The recommended temperature is half of the melting point in K for about 8 hours.

Melting

Sc may be arc or electron beam melted. Induction heating in outgassed tantalum or tungsten crucibles is most suitable. If lutetium is heated in tantalum or tungsten to temperatures significantly above its melting point, tantalum and tungsten will dissolve in the molten rare earth.

PROMETHIUM

Details

Figure 5.63 illustrates scandium's chemical symbol, element name, atomic number, and atomic weight; Figure 5.64 illustrates scandium's electron configuration.

Figure 5.65 shows a sample of promethium.

Melting point—1315 K, 1042°C, 1908°F
Phase at standard temperature and pressure (STP)—solid
Appearance—salts are pink or red color, exposed in air turns the air pale blue-green
Boiling point—3000°C

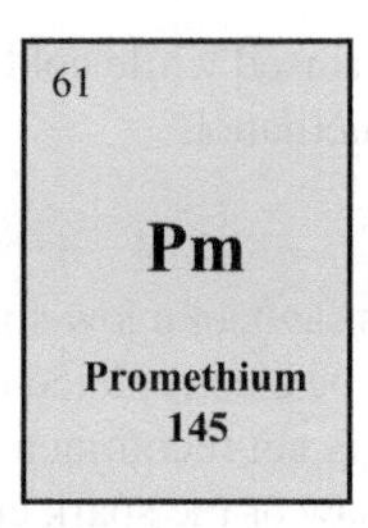

FIGURE 5.63 Chemical element designation for promethium with chemical symbol, element name, atomic number, and atomic weight.

$$\mathbf{Xe\ 4f^5\ 6s^2}$$

FIGURE 5.64 Illustrates promethium's electron configuration.

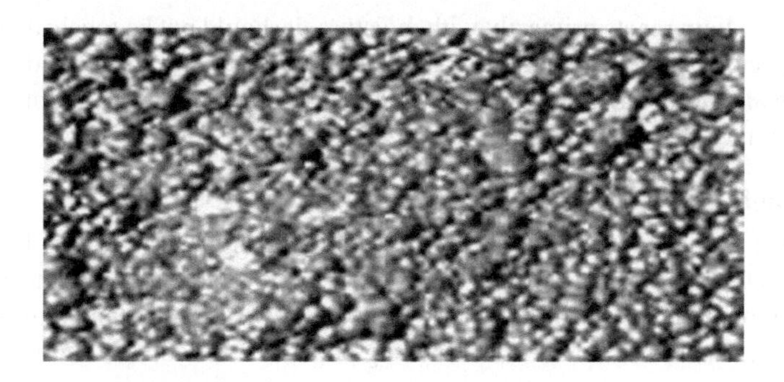

FIGURE 5.65 Promethium.

Density—7.26 g/cm^3
Solubility in water—slightly soluble
Electronegativity—1.13
Naming—after the god Prometheus who gave fire to humans

All isotopes of promethium are radioactive, toxic, the element is rare and is primarily used for research purposes in laboratories.

NOTES

1 From F. Spellman (2022). *The Science of Wind Power* (in production). Boca Raton, FL: CRC Press.
2 From Ames Laboratory SDS (2021). Washington, DC: U.S. Department of Energy.
3 From Ames Laboratory SDS (2021). Washington, DC: U.S. Department of Energy.
4 From Ames Laboratory SDS (2021). Washington, DC: US Department of Energy.
5 From Ames Laboratory SDS (2021). Washington, DC: US Department of Energy.
6 From Ames Laboratory SDS (2021). Washington, DC: US Department of Energy.
7 From Ames Laboratory SDS (2021). Washington, DC: US Department of Energy.

8 From Ames Laboratory SDS (2021). Washington, DC: US Department of Energy.
9 From Ames Laboratory SDS (2021). Washington, DC: US Department of Energy.
10 From Ames Laboratory SDS (2021). Washington, DC: US Department of Energy.
11 From Ames Laboratory SDS (2021). Washington, DC: US Department of Energy.
12 From Ames Laboratory SDS (2021). Washington, DC: US Department of Energy.
13 From Ames Laboratory SDS (2021). Washington, DC: US Department of Energy.
14 From Ames Laboratory SDS (2021). Washington, DC: US Department of Energy.
15 From Ames Laboratory SDS (2021). Washington, DC: US Department of Energy.
16 From Ames Laboratory SDS (2021). Washington, DC: US Department of Energy.

REFERENCES

Coldeway, D. (2017). Storing data in a single atom proved possible by IBM researchers. Accessed 11/08/2021 @ https://techcrunch.com/2017/03/08/storing-data-in-a-single-atom-proved-possible-by-ibm-researchers/.

Dierks, S. (2003). Dysprosium. Accessed 11/07/21 @ https://web.archive.org/web/2015 0922145520.

Emsley, J. (2001). *Samarium*. Oxford: Oxford University Press.

Emsley, J. (2002). *Nature's Building Blocks: An A-Z Guide to the Elements*. New York: Oxford University Press, pp. 181–182.

Gorman, M.R. (2009). Neodymium physical monitoring principle. Accessed 12/20/21 @ https://doi.org/10.1002/978070/162

Hajir, M. et al. (2007). Stable amorphous Calcium oxalate. Chemical Communication, May, 2007. NOTE no volume number…just May 2007.

Hammond, C.R. (2004). *Handbook of Chemistry and Physics*, 81st ed. Boca Raton, FL: CRC Press.

Hammond, J. (2005). *The Elements Handbook of Chemistry and Physics*, D.R. Lide. Boca Raton, FL: CRC Press, pp. 283–310.

Hoard, R.W., Mance, S.C., Leber, R.L., Dalder, E.N., Chaplin, M.R., Blair, K., et al. (1985). Field enhancement of a 12.5-T magnet using holmium poles. Accessed 11/08/2021 @ https://digital.library.unt.edu/ark:67531/metadc1200738.

Kamber, I., Bergman, A., Eich, A., Lusan, D., Steinbrecher, M., Hauptmann, N., et al. (2020). Self-induced spin glass state in elemental and crystalline neodymium. Accessed 10/27/2021 @ https://science.sciencemag.org/content/368/6494/eaay6757.

Krebs, R.E. (2006). *The History and Use of Our Earth's Chemical Elements: A Reference Guide*. Westport, CT: Greenwood Publishing Group.

LANL (Los Alamos National Laboratory) (2013). Scandium. Accessed 11/13/2021 @ http://periodic.lanl.gov/21.shtml.

McGill, I. (2021). Rare earth elements. Ullmann's encyclopedia of industrial chemistry. Accessed 11/11/2021 @ https://doi.org/10.1002%2F14356007.a22_607.

Lide, D.R. ed. (2005). *CRC Handbook of Chemistry and Physics*. Boca Raton, FL: CRC Press.

Radboud University (2020). New 'whirling state of matter discovered: self-induced spin glass. Accessed 10/27/2021 @ https://scietchdaily.com/new-whirling-state-of-matter-discovered-self-induced=spin-glass/.

Stwertka, A. (1996). *A Guide to the Elements*. Oxford: Oxford University Press.

Wei, Y. and Zhou, X. (1999). The effect of neodymium (Nd^{3+}) on some physiological activities in oilseed rape during calcium (Ca^{2+}) starvation. Accessed 10/25/2021 @http://www.regional. org.au/au/gcirc/2/399.htm.

6 Rare in the United States?

ARE RARE EARTHS REALLY RARE?

Okay, first off let's set the record straight: Rare earth elements (REEs) are not as rare as precious metals. Consider the price range shown in Table 6.1—we will discuss this table but for now take a look at it.

As shown in Table 6.1, the five precious metals have value because they are rare. In short, concise verbiage it can be said that in comparison, precious metals and REEs are not the same. Have you ever heard of a rare-earths rush like several historical gold rushes? Probably not. While it is true that rare earths are rarely found in nature it is also true that they are not pursued totally for their intrinsic value—that is, for ounce exchange for, in turn, cash. Instead, rare earths are sought after because they are becoming the vitamins of industry. They are important in industry because some rare earths are essential in battery construction, in permanent magnet manufacture, in use as glowing elements when electricity is passed through them and for other technical uses they are valuable—with time and further application they are becoming more valuable.

The real question is what are the estimated and inferred resources of rare earths available in the United States? This question is difficult to answer (at the present time) because with the present data at hand it cannot accurately be estimated what the reserve and resource will be available, and when, and at what rate (USGS, 2010). It can be said that the reserves and inferred resources available based on recent experience are rather thin. However, if we depend on our dependable trading partners, such as Canada and Australia, prospects for diversifying supply and meeting future demand are significantly improved. One of the negatives in mining, transporting, processing, and utilizing REEs in the United States is that it takes time to develop new rare-earth mines—we are talking about in the range of a decade or longer. The point is to forecast future supply in the future is hazardous and risky at best.

TABLE 6.1
Today's Precious Metals Prices (November 21, 2021—per ounce)

Metal	London Fix
Gold	$1,864.14
Silver	$24.78
Platinum	$1,032.00
Palladium	$2,000.00
Rhodium	$13,300.00

DOI: 10.1201/9781003350811-7

PRINCIPAL RARE EARTH ELEMENT (REE) DEPOSITS IN THE UNITED STATES

It is in the carbonatites and alkaline igneous rocks concentrated in veins genetically and spatially with alkaline igneous intrusions that the largest deposits of REEs are found. The association of REE with alkaline igneous also places REE in close association with mineral that host other valuable elements such as thorium, phosphorus, niobium, and titanium (Van Gosen et al., 2009).

So, the main question now is where are the major REE deposits found in the United States? The major U.S. REE deposits are in carbonatites and alkaline igneous complexes in veins related to alkaline intrusion, in some iron deposits associated with magmatic-hydrothermal processes and in placers (stream and beach deposits) derived from the erosion of alkaline igneous terraces.

With regard to the exact geographical location of REEs in the United States, they are present in Alaska, California, Colorado, Idaho, Illinois, Missouri, Nebraska, New Mexico, New York, Wyoming, and the Southeastern United States. Each of these regions is detailed in the following sections.

ALASKA

Well, whether it is called "the most bountiful land," or "the most beautiful country," or "the last frontier" or maybe "the land of strategic minerals and rare earth elements" Alaska is not to be ignored when it comes to its covey of critical minerals and REEs. This last statement about Alaska is quite fitting simply because Alaska is rich in critical REEs. Moreover, the characterization of critical REEs is proper when you consider that REEs are strategic minerals. Simply, Alaska is rich REEs.

With the advent of color television in the 1960s a coming out, so to speak, of the importance of REEs was brought to the forefront of attention. But this initial coming out of the importance of REEs was just the beginning of them becoming important, vital, necessary ingredients in just about every high-tech device we can think of and imagine. The truth be told we have just begun to recognize and to appreciate not only the potential uses but also the strategic value of REEs. You might be driving down the highway or Interstate and have your car/truck radio tuned into your favorite music or news station and the rich sound emanating from your earphones and not realize that part of Alaska is in your ears. And as you motor down Interstate 40 in the Texas panhandle and pass by thousands of twirling high-efficiency green-energy generators delivering needed electrical to the grid you may think it is unimaginable that the giant wind turbines contain rare-earth magnets from Alaska.

Well, one thing is certain; the United States Environmental Protection Agency (USEPA) in a 2012 and the United States Geological Survey (USGS) in a 2017 report on rare-earth minerals critical to the United States wrote or stated: "Because of their unusual physical and chemical properties, the REEs have diverse defense, energy, industrial, and military technology applications" and these statements sum-up and make clear the importance of REEs (USGS, 2017a).

Okay, what this really means is that REEs are important, critical and in the focus of those who count because the growing use, coupled with the latest news coverage about China locking up the supply chain, creating a monopoly on the worldwide supply on REEs which has raised the prominence of rare-earth elements. Although this is not a book about politics, it can't be avoided when it comes to who possesses, controls, and distributes critical resources, such as rare earths. The fact of the matter is that China has been (and continues to be) the main source of REEs for decades and since the later 1990s has accounted for more than 90% of the production of REEs—this factoid has raised concerns on a global scale. Because of China's dominance and plan to hoard REEs for its own benefits (whatever they are) the resolve of the U.S. and other global nations has focused on establishing a supply of REEs outside of China.

Okay, enough about China and its determination to dominate the world supply of REEs; that is, its attempts to corner the market on REEs. Let's shift to the United States and what is going on there in regard to the REE supply. After very little effort in discovering, mining, processing, and usage of rare earths obtained within the U.S. the United States began producing rare earths again in 2018. In resuming mining operations, the Mountain Pass mine in California was taken out of mothballs (meaning it had been under care and maintenance) and put back into operation as the only mine in operation producing this vital group of high-tech metals.

At this mine, rare earths are concentrated and then shipped overseas for further processing—leaving the United States dependent on imports for its supply of rare earth metals and compounds. Presently, roughly 80% of our rare earth metals are imported from China and the rest in smaller quantities from France, Estonia, and Japan. Because of the growing demand for REEs, along with China dominating supply the U.S. Geological Survey has listed the REEs as a member of the 35 minerals and metals considered critical to the economic wellbeing and security of the United States.

So, let's get back to Alaska again and describe/discuss REE potential within the United States' largest state. In this regard, it should be pointed out that recent work by USGS and Alaska Division of Geological and Geophysical Survey have unveiled wide swathes of Alaska that either host known deposits of REEs or are highly probable for these ever more significant constituents to contemporary gadgets, widgets, gizmos, mechanisms, and implements of national defense.

Figure 6.1 shows the location of REEs in Alaska. Many of these REE locations were well known for some time.

Bokan Mountain—Alaska[1]

For more than ten years now, Ucore Rare Metals Inc. has been endeavoring to establish a domestic source of REEs in Alaska. This work started with gearing up production at Bokan Mountain located, an area of about 3–4 mi.^{2} (7–10 km^2) in the southern area of Prince of Wales Island, Alaska, which is the southernmost island in the Alaska panhandle (see Figure 6.1).

The host rock is Upper Triassic to Middle Jurassic riebeckite–acmite-bearing peralkaline granite (i.e., a medium-to-coarse-granular acid igneous rock) with a roughly circular shape that intruded Paleozoic igneous and sedimentary rocks (De Saint-Andre et al., 1983; Lamphere et al., 1964; Staaz, 1978). The core riebeckite granite porphyry contains subordinate aplitic (fine grained, light colored) aegirine (black, blade-like structure) granite and is surrounded by an outer ring composed

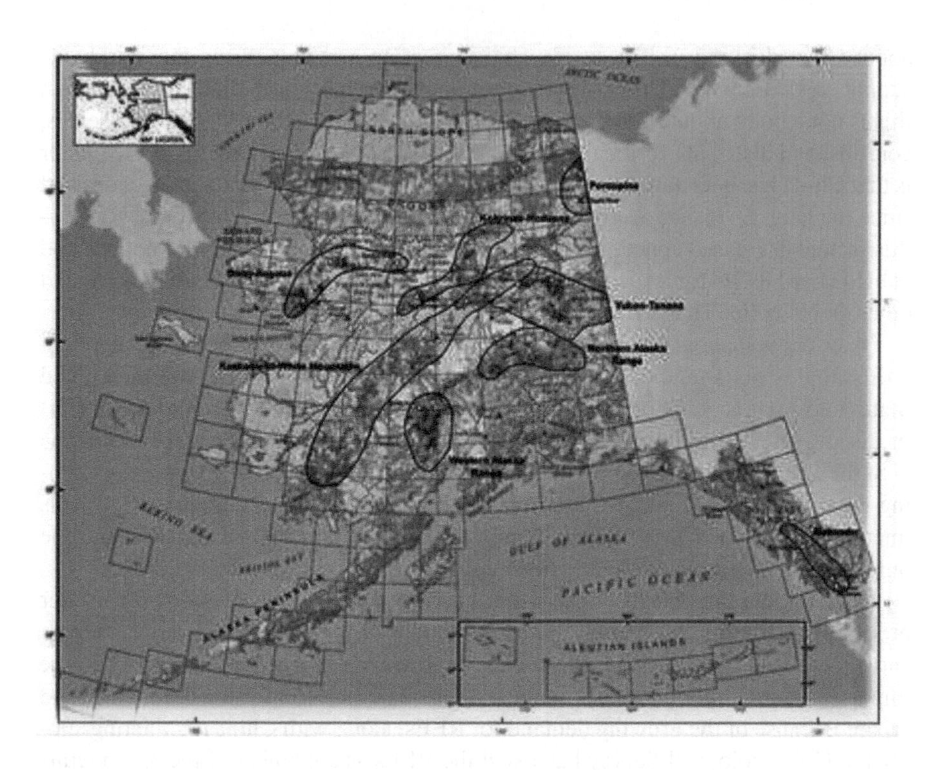

FIGURE 6.1 This figure shows the location of Alaska's known REE sites.

Source: USGS (2011). Rare elements—end use and recyclability. U.S. Dept of Interior Washington, DC.

of predominantly aegirine granite porphyry (Philpotts et al., 1998; Thompson, 1988). Emplaced in the contact zones around the intrusive granite are low levels of gold, pegmatite–aplites with thorium, and REEs (Philpotts et al., 1998; Staaz, 1978; Warner and Barker, 1989). Also, various dikes cut across all the rocks near Bokan Mountain, with configurations that include andesite, dacite, quartz, basalt, lamprophyre, rhyolite, monzonite, aplite, and quartz latite (Warner and Barker, 1989). A few of the more felsic dikes contain high levels of accessory Nb, REE, and Th.

DID YOU KNOW?

The prospects for REEs may have the greatest immediate potential for the development of REE resources in the state (USGS, 2017a).

The aplitic pegmatites (i.e., intrusive rocks) are found throughout the peralkaline granite and range in shape from lensoidal (i.e., a small geological unit that pinches out laterally) bodies to elongated or extended pods. Examples are radioactive pegmatites exposed on the east flank of Bokan Mountain, approximately 0/6 mi. (1 km) north-northwest of the Ross-Adams mine (MacKevett, 1963; Warner and Barker, 1989). Most of the pegmatites contain complex mineralogies that include zircon, albite—a plagioclase feldspar, usually pure white in color—and aegirine (crystal—see Figure 6.2), with variable amounts of fluorite, allanite, ilmenite, and riebeckite (Warner and Barker, 1963). Because of alteration of sodium-rich riebeckite, along the border zone pegmatites usually contain spread iron and hard glassy titanium spinels, as well as magnetite (one of the oxides of iron). The cores of the pegmatites consist of milky white massive quartz. The trace element compositions of the pegmatites are equally complex and may contain elevated levels of several other elements. The wall rock also contains a halo (an intrusion) that is enriched with many minerals including aegirine, sericite, and hematite alteration (Warner and Parker, 1989). Also, the feldspar within is mostly altered to clay minerals.

In 1955, uranium was discovered in the cracks, fractures, and shear zones at Bokan Mountain. Don Ross discovered the uranium with an airborne radiometric survey; later, a radioactive anomaly over the future site of the mine was prospected on the ground by Kelly Adams. Then the only active open pit mine in the area was named the Ross-Adams mine. The mine is about 0.7 mi. southeast of Bokan Mountain. The mine extracted ore from what is called the Ross-Adams pipe on the Cub claim, which lies along the contact between aegirine syenite and aegirine porphyry. In addition to uranium, Bokan Mountain contains the largest find to date of REEs in the United States. The 2,500-foot Bokan Mountain is likely to contain or host up to 4.8 million metric tons of resources including about 0.6% (63.6 million pounds) total rare earths rare earth oxides (Lasley, 2021; USGS, 2017b). Note that while the ore found at Bokan Mountain is mixed, the mix of rare earths is what makes this deposit attractive.

The U.S. Bureau of Mines investigated several prospects on Bokan Mountain and the surrounding area. They found the veins and dikes of importance because they contain anomalously high amounts of REE and other important elements.

Okay, for informational purposes, for complete understanding we should stop for a moment and take a look at what veins and dikes are or maybe at their difference.

FIGURE 6.2 Aegirine crystals. Photo by F. Spellman.

Well, the first point to make is that it is common practice to and worth noting that the terms vein and dike are sometimes used interchangeably and can actually appear quite similar, but the truth be known is that they refer to very different geologic phenomena. Dikes are a form of magmatic intrusion; they will crosscut exiting strata while other formations are concordant (conformable) with existing strata. Veins are very different; they are the result of aqueous solutions in crystalline form and occur in a wide variety of geologic settings and wide variety of climates. However, both veins and dikes are found in the same settings, around intrusions and/or adjacent to volcanic structures.

Anyway, let's get back to the dikes at Bokan Mountain. Collectively, the dikes indicate a resource of approximately 7 million tons of ore that average more than 2% REE oxides, about one-third yttrium (Warner and Barker, 1989). Additionally, the dikes are expansively enhanced in yttrium and heavy earth elements (HREE) relative to the light earths (LREE); yttrium is present at more than 1,000 times its normal crustal abundance. This composition differentiates with REE deposits elsewhere in the United States and is important because most of the HREE and yttrium are imported (Warner and Barker, 1989). On average, the dikes contain 0.7% zirconium oxide and 1.5% niobium oxide, while the amount of thorium and uranium is negligible. However, on the plenty side, in the order of generally decreasing abundance are Y, Ce, Nd, La, and Sm. Also present in noteworthy and varying concentrations are gadolinium, Dy, Ho, Er, and Tm (Warner and Barker, 1989). Note that trace to minor amounts of other valuable elements is also present, including Be, Ga, Ge, Au, Hf, Pb, Li, Pd, Rb, Ag, Sr, Ta, Sn, V, and Zn.

It is interesting to note that the minerals of the euxenite-polycrase series (aka "trash can minerals" "or the host with the most" because they are so accommodating to other minerals) host most of the Nb found in the dikes, although minor amounts of are also contained in ferrocolumbite, aeschynite, and fergusonite (rare earth oxides; Warner and Barker, 1989). Thalenite (hardness 6.5 and specific gravity 4.2), or its alteration product tengerite, contains the observed Y as well as inclusions of xenotime (a rare-earth phosphate). Other REE are contained within the minerals bastnasite, parisite, synchysite, xenotime, and monazite. Several other miners identified in the dikes include aegirine, barite, biotite, calcite, epidote, fluorite, galena, iron oxides, magnetite, microcline, microperthite, native silver, pyrite, riebeckite, sphalerite, and zircon (USGS, 2010). Philpotts et al. (1998) in a USGS report dealing with petrogenesis of late-stage granites and Y–REE–Zr–Nb-enriched vein dikes of the Bokan Mountain stock examined a pers 3-km (1.9 mi.) transect from the margin of Bokan Mountain peralkaline granite stock along a micro-pegmatite and aplite vein-dike system enriched in Y–REE–Zr–Nb, and they identified minerals such as arfvedsonite, tainiolite, and gittinsite, as well as several other REE, –Zr–Nb-bearing phases. Philpotts et al. (1998) point out that textures range from primary igneous assemblages, through autometasomatic replacements [i.e., a metamorphic process by which the chemical composition of a rock or rock portion is altered in a pervasive manner and which involves the introduction and/or removal of chemical components as a result of the interaction of the rock with aqueous fluids or solutions. During metasomatism the rock remains in a solids state (Zharikov et al., 2007), to late-stage fracture fillings of hydrothermal aegirine, zircon, iron oxide, iron hydroxide, and clay minerals.

Philpotts et al., also reported that inclusions are abundant. They went on to report that mineral rare-earth abundance shows diversity that contrasts with the limited range of whole-rock patterns. Yttrium and heavy rare earths are unusually enriched in these samples. The rocks display a pronounced negative europium anomaly (Eu/Eu* = 0.27 ± 0.03) that is consistent with their sampling of the same evolutionary stage of an igneous system that has undergone extensive feldspathic fractionation. The enrichment of trace elements in the vein dikes likely resulted from the separation of solid, fluid, and vapor phases. Included is evidence for immiscible fluid separation include silicate-, fluoride-, and sulfate-phases. Measurements for neodymium isotope indicate depleted source rocks for the magma and subsequent open geochemical systems (Philpotts et al. 1998). By use of various analytical methods, the examined transect was found to be generally enriched in Y and HREE and to have a pronounced negative Eu anomaly, which largely agrees with the results obtained by Warner and Barker (1989) for several dike systems in the Bokan Mountain area.

SIDEBAR 6.1 WHEN A GEODUCK IS NOT A GEODUCK

Geoduck? What does the famous clam, geoduck, the one that made Seattle Public Market famous and shown in Figure 6.3 have to do with rare earths? Well, nothing. Not a thing. However, the Geoduck Prospect of Bokan Mountain is definitely apropos to our discussion of REE in Alaska.

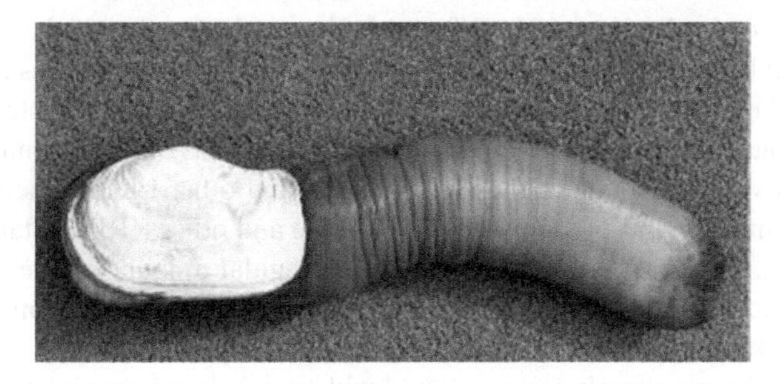

FIGURE 6.3 Photo of Geoduck clam taken in Seattle from The Public Market (2007).

The Geoduck Prospect of Bokan Mountain is a dike type deposit, an occurrence that is related to systems of equigranular, fine- to medium-grained, andesite dikes approximately 2,900 m in length with a width of approximately 0.5 m at an estimated depth of 760 m. The dikes have steep to vertical dips, cut Silurian and Ordovician quartz monzonite, granite, and quartz diorite, and typically have wall rocks marked by chlorite and epidote alteration (MacKevett, 1963; Warner and Barker, 1989). The ore mineralogy is still being tested and evaluated in detail but it is probably similar to that in the mineralized dikes at the Dotson Prospects (a shear zone, fractured controlled deposit type), which may be extensions of the Geoduck dikes. Many samples contain elevated values of beryllium, thorium, yttrium, REE, columbium, uranium, and zirconium.

Ucore Uranium (2010) in 2009 drilled two holes near the geoduck prospect for rare-earth elements. The holes are along the "Geoduck trend: that is oriented northwest along the andesite dike". The analyses for the samples were reported as the light rare-ear-element oxides or LREO (lanthanum, cerium, praseodymium, neodymium, and samarium) and the heavy rare-earth-element oxides or HREO (europium, gadolinium, terbium, dysprosium, holmium, erbium, thulium, ytterbium, lutetium, and yttrium); together they are termed the TREO, the total rare-earth-element oxides. The most notable intercept in the two holes was 0.4 meter with 3.4% TREO (USGS, 2010). The ratio of the HREO to the TREO in the three best intercepts varied between 41.9 and 46.6%; in other words, nearly half of the rare-earth elements in the samples are the heavy rare-earth elements.

With regard to the other dike systems (i.e., the Cheri, Upper Cheri, Dotson, Geiger, I & L vein system dikes; see Figure 6.4) the minerology of the I & L vein system is complex. Specifically, U. Th, and REE are present in several different minerals from different parts of the veins, and they are usually in gangue (i.e., in valueless material in which ore is found) dominated parts of the veins, and they are usually in a gangue dominated by quartz and albite (Keyser and McKenney, 2007; Staaz, 1978). Generally, U is located in Th-bearing uraninite, although it can be found in brannerite in some of the transverse veins (Staaz, 1978). Kasolite and Sklodowskite, secondary U minerals have been identified, but they are rare (Keyser and McKenny, 2007). Thorite is the main thorium mineral found in the northwest part of the vein system, while allanite is found in the southwestern part of the system as well as in the transverse veins. The other rare earth minerals besides allanite include bastnasite, xenotime, and monazite. However, the distribution of the REE oxides is unusual in that one part of a vein may contain mostly bastnasite and other LREE whereas another part of the same vein has predominantly xenotime and other HREE (Staaz, 1978).

In addition, the I & L vein system contains irregular amounts of Be, Nb, Zr, Ba, Sr, Sn, Cu, and Mo. Most of the Nb-rich minerals are located in euxenite-polycrase or ferrocolumbite-ferrotantalite series.

Regularly, these minerals are associated with zircon as microveinlets. Otherwise, these minerals may replace albite and quartz in the vein matrix. Other minerals include aegirine, barite, biotite, calcite, epidote, fluorite, galena, iron oxides, potassium feldspar, magnetite, pyrite, native silver, riebeckite, and sphalerite (Staaz, 1978).

Note that many of the dikes in the Bokan Mountain area crop out in heavily forested areas and that the bedrock is covered by thick vegetation (Warner and Barker, 1989). For instance, the Upper Cheri dike system (366 m in length and 1 m wide) can be traced only intermittingly because of muskeg cover, glacial till, or obscuration by hillside talus where it is open-ended. On the northwest end of the upper Cheri, the dike system passes under a low-lying creek valley (Warner and Barker, 1989). Detailed mapping in the I & L vein system is prohibited/hindered by heavy soil cover. Also, accessibility to the area is limited given that the topography of the Bokan Mountain ranges from relatively steep to precipitous (Keyser and McKenney, 2007). Although deep-water marine access is available to and from Ketchikan and

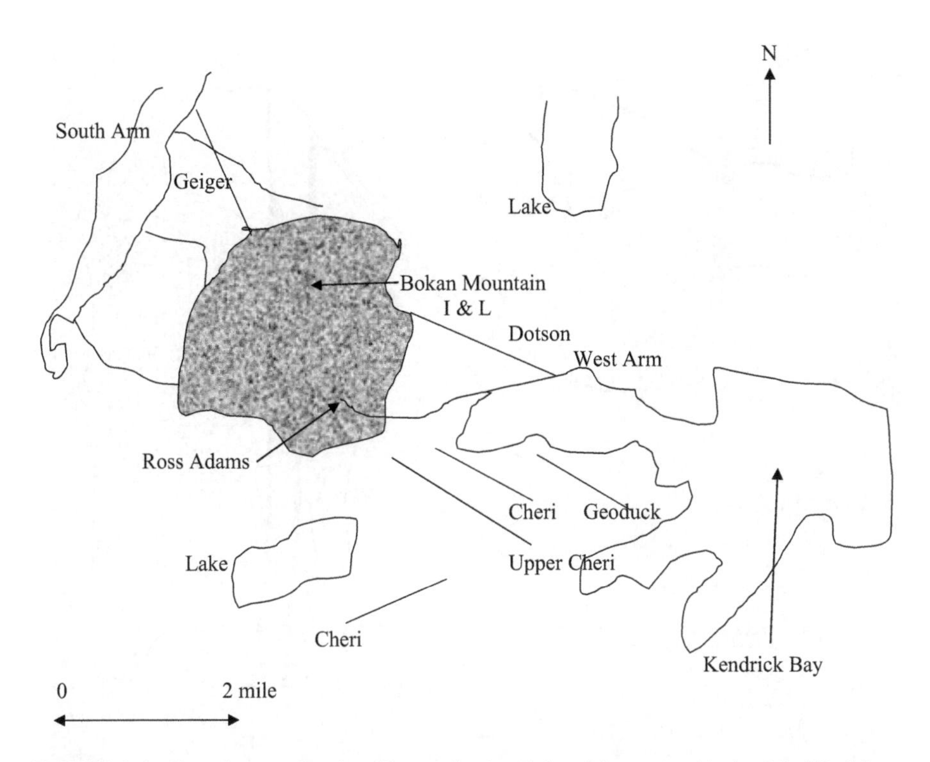

FIGURE 6.4 Rough map of major dikes and veins Bokan Mountain, Alaska. Modified from Heylmun (1999).

Prince Rupert by way of Kendrick Bay and Moira Sound, accessing more remote sections of the area must be achieved by foot, boat or helicopter. In spite of these limitations, the mineralization could have considerable economic potential.

Salmon Bay, Alaska

Salmon Bay is located on the northeast shore of Prince Wales Island, the southernmost island in Alaska. In 1984 and 1985, the Bureau of Mines investigate radioactive carbonate veins near Salmon Bay for concentrations of columbium and associated metals (see Figure 6.5). The veins cut units of graywacke, conglomerate, argillite, and limestone and range in width from less than an inch to great that 10 ft and have a length ranging from less than a hundred to greater than 1,000 ft.

The predominant minerals in the veins are dolomite–ankerite and alkali feldspar, with lesser amounts of hematite, pyrite, siderite, magnetite, quartz, chalcedony, and chlorite (Houston et al., 1955). Other minerals identified include parisite, bastnasite, muscovite, fluorite, apatite, thorite, zircon, monazite, epidote, topaz, garnet, chalcopyrite, and marcasite. It is the thorite and monazite in the veins that cause the radioactivity; they both contain thorium. The fluorcarbonates parisite and bastnasite are found in nonradioactive carbonate–hematite veins, located along the coast, and are wider than radioactive veins. Houston et al. (1955) point out that the two fluorcarbonates, parisite is more abundant and appears to be a late-stage mineral that fills

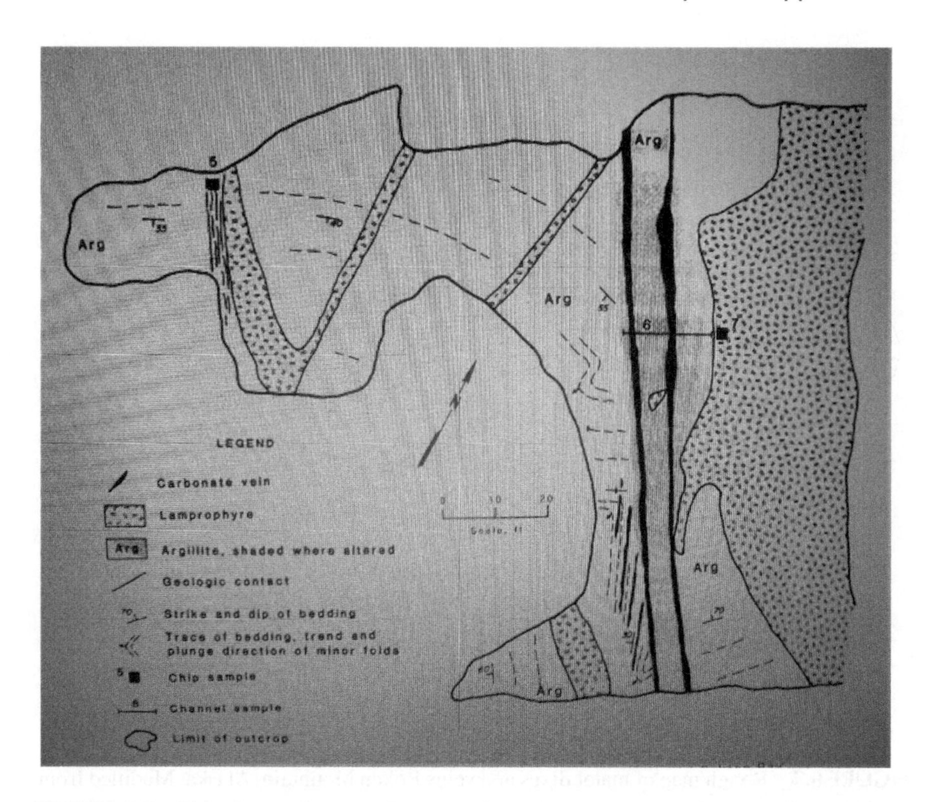

FIGURE 6.5 This figure shows carbonate vein and lamprophyre dikes in area north of Salmon Bay.

Source: US Department of Interior (1989).

in small vugs (i.e., a small cavity in a rock filed or line with crystals different from the host rock) or was deposited along fractures in the host carbonate vein (Houston et al., 1955). Rock sampling indicates that only three veins contain columbium, REE, or thorium mineralization over significant strike lengths. Gravel samples collected from the beach and offshore indicate that some of the sediments near Salmon Bay are slightly enriched in columbium, uranium, and thorium; the gravel sample grades are too low to be considered resources (US Department of Interior, 1989).

The bottom line: Additional exploration and analysis of the Salmon Bay deposit are necessary to determine if there is any economic value of the resource. To date, little work has been done, but Salmon Bay has been identified as a REE prospect.

California

Mountain Pass Area

It can be said that discovery, transition, evolution, and change are normal occurrences in life—we can say it is nothing unusual, is expected, and is simply modulation. Well, this set of evolutionary circumstances certainly is applicable to discovery, transition, evolution, and change that has occurred in the world of REE deposits in the Southeast Mojave Desert. Specifically, it is the Mountain Pass region of California that is being referred to as our sample of the discovery of, transition of, evolution, and change of pertinent to our discussion herein.

Okay, let's put this train of discussion into the simplest terms possible via a basic explanation. First, rare earth minerals were discovered at Mountain Pass in 1949. Second, REE was mined and processed for profit, but not enough financial gain could be made from the initial project—thus, bankruptcy occurred. Third, the technological world started using REE for critical applications in defense, then wind turbines, flat screen television, and then there is that push to substitute fossil fuel-powered vehicles with electric vehicles (EVs). Fourth, because REE has found its new niche, so to speak, and is highly desirable and demanded, mining, and processing REE has become a major goal of the United States. Now, this is especially the case when we presently depend on the supply of REE from China and other nations.

So, someone woke up and asked the question: Why?

This is absolutely a good question when you consider that we have our own resources in our own country. And one of those resource locations is at Mountain Pass, California and is part of the Clarke Mountain Range and is now being actively mined for REE (see Figures 6.6 and 6.7). Mountain Pass has a massive carbonatite called the Sulphide Queen and hosts of REE ore body in the district. The carbonatite ore body has an overall length of 730 m (2,400 ft) and an average width of 1220 m (394 ft). The typical ore contains about 10–15% ore mineral bastnasite, 65% calcite or dolomite (or both), and 22% barite, plus other accessory minerals (Castor and

FIGURE 6.6 Clarke Mountain Range. Photo by K.M. Denton USGS (public domain).

FIGURE 6.7 Mountain Pass REE mining operations.

Source: USGS Bradley Van Gordon (2012) (public domain).

Nason, 2004). Note that Sulphide Queen carbonatite body is the largest known (the key word is "known") mass of REE ore in the United States.

The REE content of Mountain Pass was first discovered in 1949 by a couple of uranium prospectors recorded modest radioactivity on their Geiger counter on Sulphide Queen hill and also at what is called the Birthday vein, located 1200 m (4,000 ft) to the northwest (radioactivity reflected the thorium content of the carbonatite). After taking samples of radioactive rock, they took them to the U.S Bureau of Mines office in Boulder City, Nevada. Analyses of the samples by the Bureau of Mines, confirmed by laboratories of the USGS, found that the rock was rich in bastnasite—a rare earth–carbonate–fluorine mineral—which subsequently became the primary ore mineral of Mountain Pass. Later in 1949, the prospectors filed claims on the Birthday vein system. In November of 1949, USGS initiated what they classified as a high-priority field study of the Mountain Pass district. This study mapped, described, and the district in detail; it was this study that led to the discovery of the massive Sulphide Queen carbonatite stock (Olson et al., 1954; Shawe, 1953).

DID YOU KNOW?

Carbonatite is often confused with marble and only geochemical verification can be used to verify; that is, for the uninitiated. It is a type of intrusive or extrusive igneous rock defined or classified by mineralogic composition consisting of greater than 50% carbonate minerals (Bell, 1989).

You may be wondering about a more than a half-century old report by several of the cited authors herein as being pertinent to the present time. Well, the truth be known their individual reports remain the most comprehensive published account on the geology of the Mountain Pass district. As the old saying goes the proof is in the pudding and in this case the geologic mapping, lithologic descriptions, mineralogy described in their report have been proven through the several decades of development to be an accurate accounting of the district. Castor (2008) detailed more information resulting from recent geological research in the district, including subsurface information.

Music Valley, California

Just northeast of Joshua Tree National Park and about 10 mi. (16 km) southeast of Twenty Palms in Riverside County lies Music Valley in the Pinto Mountains. A reported xenotime deposit is located in the Music Valley area situated in the Pinto Gneiss estimated to be of Precambrian age. Biotite-rich zones in the gneiss can contain abundant orange xenotime grains (USGS, 2010).

DID YOU KNOW?

The geological history of the Music Valley area and formation of the Pinto Gneiss spans an undetermined length of time and involves a long sequence of intrusions and periods of metamorphism. Probably the first event which can now be deciphered was the sedimentary and possible volcanic accumulation of material that has since been metamorphosed to form the Pinto Gneiss (California Division of Mines, 2007).

With regard to the production, the Music Valley prospect has received little activity and no current exploration interest. Small-scale exploration of these deposits the 1950s investigated their radioactivity.

SIDEBAR 6.2 WHEN SALT TURNS GOLDEN?

In the beginning, it was an accidental creation.

It has been called a mistake, a slip-up, a gaffe, a blunder, a total mess. Well, about a year ago when I walked several miles around its southern extremity in 105-degree heat on a cloudless day my impression was that it was more of a garbage dump than anything else. I also walked around what is left a planned and developed resort community that is now a ghost town—paradise lost, one could say, with mostly empty lots and demolished homes—looks like a post-apocalyptic ghost town, on steroids. I also discovered that there are few things in life uglier than palm tree remnants, stumps. And the land-locked sea so beautiful, but yet so toxic. Well, hells bells, it is a matter of or could be the results of when humans step in and foul things up that there are consequences.

Consequences?

Yes. For instance, the location being referred to herein is a fish and bird killer. It is a location where the disease has set in and wildlife has taken the toll; a life-ending levy. It is a body of water that is shrinking because agricultural runoff into the water body started to shrink (and still is) when agricultural practices became more efficient. Oh, then came the drought and the accompanying drying out and worse, the winds—the incessant winds—that can carry cold and hot and dust. And with the lack of moisture, the drying of the sea and then the winds the dust rises, becomes airborne and adds to the smog-laden air of Los Angeles, and other locations.

While surveying part of the remnants of the place I happened to run into a vagrant, a friendly guy named Joe. He told me that he remembered the glory days of the place when visitors used to fish, swim, and sail their boats from what are now crumbling docks on dry land that used to bridge water. "Them folks were in paradise that soon turned to the nether world, hell on earth." This is what Joe told me as he wandered off along the receding shore soaking up more of than burning sun; where he was headed, I had no clue.

Anyway, I have painted a bad picture of the shallow, landlocked, highly saline body of the Salton Sea in California (see Figure 6.8). But it must be said that there are two sides to each coin, one might be negative and the other positive, so to speak. And in this case, there certainly is.

FIGURE 6.8 Salton Sea, California.

Source: USGS public domain photo (2019). Accessed 12/7/2021 @ https://www.usgs.gov.media/images/saltonsea.

The positive?

Deep below the Salton Sea is a treasure trove of lithium. You know, the element used in manufacturing the best batteries available at the present time. Mining for lithium can be considered a modern gold rush, of sorts. The "gold" is the white mineral lithium that is removed from the geothermal brine at the Salton Sea. Now it is important to point out, even though it has been said that the salt of the Salton Sea turns to gold this is not quite true—the lithium does not come from the salt in the Salton Sea, this is simply a misconception. Instead, the lithium comes from the superheated fluid (brine) deep below and the extraction procedure is a geothermal process.

So, are we saying that lithium is a rare earth element? No. Lithium is not a rare earth element but many think it is. So, why are we discussing lithium? Good question. And the answer provided herein is to set the record straight.

Lithium is not a rare earth mineral. But it certainly is a critical resource just as REE is. I have included lithium herein to make a point and point is that we need to mine, process, and produce our own natural resources to the hilt—and to depend on no one else for supplies. Whether that be lithium or REE.

Colorado

Iron Hill Carbonate Complex

The Iron Hill Carbonate Complex is located near the small town of Powderhorn, about 22 mi. (35 km) south–southwest of Gunnison, Colorado. A massive carbonate stock forms the core of Iron Hill carbonatite complex. Carbonatites are rare intrusive or extrusive carbonate igneous rocks that are formed by magmatic or metasomatic processes. Carbonatites generally consist of 50% or more (by volume) primary carbonate minerals, such as calcite, dolomite, and (or) ankerite. They are genetically associated with, and therefore typically occur near, alkaline igneous rocks. The alteration process with in carbonatites prevents the formation of carbonate minerals; instead, carbonatites crystalize from a magma that is super-saturated in calcium and carbon dioxide. When it comes to attempting to explain the genesis of carbonatites several theories have been put forth. Probably the most accepted theory is that carbonatites were formed due to the partial melting of peridotites that make up much of the Earth's crust.

Carbonatites, as mentioned earlier, generally occur near alkaline igneous rocks—sometimes referred to as "alkalic rocks"—are a series of rocks formed from magmas and fluids so enriched in alkalies that sodium- and potassium-bearing minerals form constituents of the crock in much greater proportion than in "normal" igneous rocks.

Van Gosen and Lowers (2007) reported that no minerals have been produced from this intrusive complex. Its interests have focused on the substantial titanium resource within the pyroxenite unit of the complex. Currently (2021), it appears that no minerals have been produced from the Iron Hill Carbonatite Complex.

Wet Mountains Area, Colorado

The Wet Mountains Area of Colorado is located in Fremont and Custer Counties of south-central Colorado. Armbrustmacher (1988) notes that mineral deposits in the area include REE and thorium in veins, dikes, fracture zones, and carbonatite dikes associated with three Cambrian alkaline complexes (Olson et al., 1977) that intruded the surround Precambrian terrane. The fracture zones and thorium–REE-mineralized veins and fracture zones, distal to (i.e., the outer part of area affected by geological activity) the three alkaline intrusive complexes, have the highest economic potential for REE and thorium resources. Note that the REE–thorium veins and fractures zones are linear features, typically 3.3–6.6 ft (1–2 m) thick, but a few are almost 50 ft (15 m) thick. A few individual thorium veins can be traced in outcrop up to 1 mi. (1.5 km) and some radioactive fracture zones as much as 8 mi. (13 km). Christman et al. (1953, 1959) mapped nearly 400 veins in this area with most of these veins within a 22 mi.^{2} (57 km^2) tract of Precambrian gneiss migmatite located southeast of a quartz syenite complex at Democrat Creek. Many of the prospective vein and fracture-zone deposits occur on private lands. No apparent exploration activity is underway at present (2021).

Idaho

Diamond Creek Area, Idaho

On the eastern slope of the Salmon Mountains, the vein of interest here is located and it is 8 mi. (13 km) northwest of Salmon, Idaho. Veins of the Diamond Creek district are found throughout the area 2.5 mi. (4 km) long by 0.5 mi. (0.8 km) wide. Note that the veins are hosted by Proterozoic quartzite (later part of the Precambrian era when the earliest forms of life evolved) and siltite and by Mesoproterozoic granite. The Diamond Creek veins are mineral fillings in fractured and sheared bedrock; the veins are up to 25 ft (7.6 m) thick in the metasedimentary rocks (quartz and siltite) but rarely more 2 ft (0.6 m) thick in the granite (Staaz et al., 1979). These veins contain considerable amounts of hydrous iron oxide minerals, accompanied by disseminated thorium—REEs-bearing minerals. The vein deposits with copious amounts of yellow-to-brown iron oxides (limonite and goethite) appear to contain the highest thorium (Th) and REEs concentrations (USGS, 2010). In regard to the present status of the hunt for and processing of REE not much has been accomplished up until recently (2020) when New Jersey Mining Company acquired the REE deposit which consists of 427 hectares (1,041 acres) located In Eureka Mining district 8 mi. (13 km) north–northeast of Salmon, Idaho (Long and Van Gosen, 2010).

Hall Mountain, Idaho

On Hall Mountain in northernmost Idaho—1 mi. (1.6 km) south of the United States–Canada border—veins crop out in an area about ½ mi.^{2} (2.6 km^2)—6,000 ft (1,830 m) by 1,000 ft (305 m) wide. The veins cut Precambrian quartzite and quartz diorite. In exposed length, the veins range 6 ft (1.8 m) to 700 ft (213 m) and in width from 0.6 ft (0.18 m) to 13 ft (4 m). The primary thorium- and REEs-mineral is thorite. The most abundant gangue minerals are quartz and calcite, associated with chlorite magnetite, limonite, pyrite, and biotite, along with numerous minor and trace elements. A total of 30 minerals were identified by Staaz (1972a). Currently (2021), no active exploration has been reported in the district.

Lemhi Pass District—Idaho–Montana

Straddling the Continental Divide on the Montana–Idaho border in the central Beaverhead Mountains is the Lemhi Pass District which contains numerous vein deposits of enriched thorium and REEs within a 54 mi.^{2} (140 km^2) core area. The core area is approximately 154 mi.^{2} (400 km^2). Staaz (1972b; Staaz et al., 1979) while in the Lemhi Pass District mapped 219 veins enriched with thorium and REE. Most of these veins are quartz–hematite–thorite veins that fill shears, brecciated zones, and fractures in Mesoproterozoic quartzite and siltite host rocks. Thorium and REE also appear in monazite–thorite–apatite shears and in replacements with biotite, alkali feldspar, and specularite. The thorium–REE veins of the district range from 3.3 ft (1 m) to at least 4,347 ft (1,325 m) in length and from a few centimeters (1 in.) to as much as 12 m (39 ft) in width. The longest and widest vein in the district—the Last Chance vein—is 4,348 ft (1,325 m) long and 10–26 ft (3–8 m) wide for most of its length (Long and Van Gosen, 2010). Currently (2021), no thorium and REE have been produced from this district. Past exploration, sampling, testing, and development in these veins focused on their thorium content; previous development involved trenching several veins which produced modest underground workings in the Last Chance vein.

Illinois

Hicks Dome

The dome-shaped structure known as Hicks Dome is located in Hardin County, southern-most Illinois. The dome is 9 mi. (14.5 km) in diameter and was formed by the displacement of sedimentary rocks at least 3,900 ft (1,200 m) upward above an alkaline intrusion in depth. An explosive intrusion of magma pushed up 1,979 ft (600 m) of limestone (mostly). A hole drilled near apex of the dome (Brown et al., 1954) intersected a mineralized breccia at a depth of 1,600 ft (490 m), which continues to the bottom of the hole at 2,940 ft (897 m). Thorium and REE are contained in mineralized breccia, tentatively identified as residing in monazite, and are found in association with florencite, a cerium–aluminum phosphate; gangue minerals are felspar, calcite, quartz, minor pyrite, and traces of sphalerite and galena (Long and Van Gosen, 2010). Currently (2021), no active exploration is underway.

Missouri

Pea Ridge Iron Deposit and Mine

The Pea Ridge iron orebody and mine are located about 60 mi. (97 km) southwest of St. Louis in Washington County. The breccia pipes that cut through the Pea Ridge massive magnetite–iron ore body. The Precambrian volcanic rocks of the St. Francois terrane of southeastern Missouri are the host of the Pea Ridge deposit.

The magnetite-rich orebody is interpreted as a high-temperature, magmatic-hydrothermal deposit (Sidder et al., 1993) in ash flow tuffs and lavas, which may have formed in the root of the volcanic caldera (Nuelle et al., 1991). Four mapped breccia pipes contain REE that steeply crosscut the magnetite–hematite orebody and its altered rhyolite host rock. Exposed portions of the breccia pipes are as much as 197 ft (60 m) in horizontal length and as much as 49 ft (15 m) in width; below the mined levels contain pipe extensions of unknown length (Seeger et al., 2001). REEs-bearing minerals in the breccia pipes include monazite, xenotime, minor bastnasite, and britholite. Although the REEs reported in breccia pipes are consistently high but variable. Nuelle et al. (1992, p. A1) state, "Total REE oxide content samples of the groundmass material, which are not diluted with lithic fragments, average about 20 weight percent." Seeger et al. (2001, p. 2) state, "Total REE oxide concentrations of grab samples range from about 2.5 to 19 weight percent." Bulk sampling by the Bureau of Mines found REE oxides concentrations ranging from 7 to 25 weight percent and an average of 12% (Vierrether and Cornell, 1993). Currently (2021), there is no active development in this deposit.

Nebraska

Elk Creek Carbonatite

In the subsurface southwest of the small town of Elk Creek in southeastern Nebraska lies a buried REEs- and niobium (Nb)-rich carbonatite mass. Based on exploration drilling and the extent of magnetic and gravity anomalies, the entire oval-shaped, subsurface body, which is recognized by a geophysical anomaly caused by the carbonatite and associated intrusive rocks, is about 4.3 mi. (7 km) in diameter. Analyses of drill core showed the intrusion at depth comprised mostly massive to brecciated, apatite- and pyrochlore-bearing dolomitic carbonate (89%), along with fenitized (i.e., metasomatic alerted) basalt, lamprophyre, and syenite (totaling 11%). Major-element analyses suggest that the carbonate mass is a magnesian carbonatite (dolomitic), generally similar in gross chemical composition to the Iron Hill (Powderhorn) carbonatite stock in southwestern Colorado (Long and Van Gosen, 2010). The REE is hosted principally by the minerals bastnasite, parisite, and synchysite by smaller amounts of monazite (Xu, 1996). Niobium was deposited in pyrochlore. The U.S. Geological Survey obtained a potassium–argon age on biotite in the carbonatite of 544 ± 7 million years old (Xu, 1996). No mineral resources have been produced from this intrusion (Long and Van Gosen, 2010).

New Mexico

Note: New Mexico has some deposits that are in the early exploration stage and it will take years for these deposits to be developed, if they are economic. Because world-wide demand for REE is increasing daily it is important to understand the REE potential in New Mexico, even if the deposits are not produced in the next few years, because these resources are important for future technological development critical to maintaining our standard of living and for defense purposes (McLemore, 2015).

Capitan Mountains

The Capitan deep-seated intrusion of igneous rock (a pluton), which lies along the east-west trending Capitan lineament (Allen and Foord, 1991; Allen and McLemore, 1991), is the largest exposed Tertiary intrusion (outcrop area of approximately 280 mi.^{2}) in New Mexico. The best estimate of the age of the pluton is 28.8 million years ago (28.8 Ma), based on $^{40}Ar/^{39}Ar$ dating (this is a more sophisticated variation of the K/Ar dating techniques—$^{40}Ar/39^{39}$ is used as the parent) of adularia that is associated with emplacement of the pluton (Dunbar et al., 1996). Thorium and REE resources in this district have not been fully estimated. Staaz (1974) analyzed 17 samples in the veins and found thorium of less than 0.01 to as much as 1.12%; however, some assayed vein material showed as much as 1.7% thorium. No mineral resources have been produced from these vein deposits (Long and Van Gosen, 2010).

El Porvenir

The El Porvenir or Hermit Mountain district lies about 15 mi. (24 km) northwest of Las Vegas, New Mexico and 3 mi. (4.8 km) north of Porvenir, on the eastern edge of the Las Vegas Range, San Miguel County, north-central, New Mexico. The largest part of Hermit Mountain is formed by a pink, coarse-grained Precambrian granite that is cut by pegmatite (igneous rocks) dikes and quartz veins (Robertson, 1976). Some of the pegmatites reportedly contain monazite and REEs mineralization (Long and Van Gosen, 2010). Presently (2021), little published information is available on the chemistry of these pegmatites, but the data that are available suggest that anomalous rare elements are present; however, no mineral resources have been produced from these occurrences.

Gallinas Mountains

Gallinas mining district lies in the Gallinas Mountains and is one of several mining districts associated with alkaline igneous rocks in the Lincoln County porphyry belt in central New Mexico, about 10 mi. (16 km) west of the town of Corona. The cerium-rich mineral bastnasite precipitated in fluorite-copper sulfide deposits in the Gallinas Mountains. The mineral-resource potential is high with a moderate level of certainty for REE; moderate with a moderate level of certainty for gold, silver, and iron; and low with a moderate level of certainty for copper, molybdenum, uranium, and tellurium in the Gallinas Mountains district (Long and Van Gosen, 2010; McLemore, 2018). Presently (2021), there appears to be no active exploration in this district.

Gold Hill Area and White Signal District

The White Signal district is located in Grant County, southwestern New Mexico and at the great of the Burro Mountains lies the adjacent Gold Hill area. Pods and lenses of REEs–thorium-bearing minerals within pegmatites are hosted by Proterozoic Burro Mountain granite in the western part of the White Signal district. The primary minerals of these pegmatites are quartz, muscovite, and microcline. Locally, large euhedral crystals of euxenite (meaning it is bounded by faces corresponding to its regular crystal form unconstrained by adjacent minerals) are found locally, and some crystals are several inches long (Gillerman, 1964). Other REE-bearing minerals reported in the pegmatites are allanite and samarskite (Richter et al., 1986). In the Gold Hill area, near the crest of Burro Mountains, the same REE-bearing minerals are hosted in similar but larger pegmatites that also cut the granite in Burro Mountain (Hedlund, 1978). Milky quartz, microcline, albite, and muscovite are the primary minerals and allanite, euxenite, and samarskite are the REE-bearing minerals (Long and Van Gosen, 2010). Currently (2021), there is no active exploration or development in the district.

Laughlin Peak Area

Laughlin Peak prospects is a 7300 ft REE, niobium and thorium mine located in Colfax County, New Mexico. In this district, Staaz (1985) mapped 19 veins ranging from 1.6 to 1,800 ft (0.5 to 550 m) in length 0.08 28 in. (0.2 to 70 cm) in thickness. REE-bearing and thorium mineral in the veins include brockite, xenotime, and crandallite. The brockite and xenotime are mainly enriched in the yttrium-group (heavy) rare earths, whereas the crandallite contains mostly cerium-group (light) REEs. The veins are steeply dipping and lie along fracture zones, cutting intrusive breccia and trachyandesite but mostly trachyte and Dakota Sandstone. The gangue minerals are mostly potassium feldspar, quartz, or calcite, and lesser amounts of goethite, magnetite, barite, zircon, rutile, and a manganese oxide (Long and Van Gosen, 2010). There has been some prospecting for radioactive deposits but currently (2021) there appears to be no active exploration in this district.

Lemitar and Chupadera Mountains

Lemitar Mountains in west-central New Mexico contain carbonatite veins and dikes, and more than a dozen similar carbonatite dikes are known to be located in the south within the adjacent Chupadera Mountains. These mountain ranges lie west of San Antonio, Socorro and Lemitar in Socorro County, New Mexico. With regard to the basic geology and deposit types more than 100 carbonatite veins and dikes that contain REE cut Precambrian metamorphic and granitic terrane in the Lemitar Mountains, and more than dozen similar carbonatite dikes intruded Precambrian metamorphic rocks to the south within the adjacent Chupadera Mountains (McLemore, 1983a, 1987; Van Allen et al., 1986). The carbonatite intrusions (veins) range from less than 0.4 in. (1 cm) thick to more than 3.3 ft (1 m) thick dikes (McLemore, 1983a, 1987). A few of the dikes can be traced in outcrop for as much as 1,900 ft (600 m). Note that dike swarms are formed in subparallel sets of carbonatites locally. Alkaline igneous rocks are lacking in these mountains, so their igneous source presumably lies at some depth (McLemore, 1987). Using the potassium–argon aging method suggests that they are Ordovician (449 ± 16 million years old, McLemore, 1987) and thus represent a part of widespread Cambrian–Ordovician igneous activity in New Mexico (Long and Van Gosen, 2010). Currently (2021), there appears to be no active exploration in this district.

Petaca District

The Petaca District lies west of the Rio Grande near the eastern margin of Rio Arriba County, in rugged, forested of the Tusas Range. The district is accessed by US Forest

Service roads whose condition ranges from good to poor and unpaved. Thorium and REE-bearing pegmatites crop out in Precambrian rocks in the southeastern Tusas Mountains (Bingler, 1968). The pegmatites (which were mined for scrap and sheet mica for several years) of the Petaca district take an assortment of shapes, such as pipes, dikes, sills, pods, troughs, and irregular forms. These pegmatites crop out for 75 to 1,430 ft (23 to 436 m) in length and have an average width 30 to 35 ft (9 to 11 m) (Bingler, 1968). A REE elevated concentration within the pegmatites consists mainly of the mineral samarskite, a REE–iron–uranium–thorium–niobium–tantalum–titanium-bearing oxide (Long and Van Gosen, 2010). Currently (2021), there appears to be no active exploration in this district.

Red Hills Area

In the Red Hills area of the southern Caballo Mountains, Sierra County, New Mexico dike-like and tabular bodies containing thorium and REEs are present and exposed. Alkaline igneous complexes similar to carbonatites and metasomatic rocks found in New Mexico are well known for potential economic deposits of REE, U, Nb, Zr, Hf, Ga, and other elements (Long and Van Gosen, 2010; McLemore et al., 2012). K-feldspar-rich, brick-red rocks (deep red syenite), called episyenites are present containing anomalous concentrations of REE, U, Th, and other elements and crop out across an area 3 mi.^{2} (7.8 m^2), centered about 2.5 mi. (4 km) from Caballo dam (Long and Van Gosen, 2010). Currently (2021), no resources have been produced from these outcrops and there appears to be no active exploration in the district.

DID YOU KNOW?

The term episyenite is used to altered rocks that were desilicated and metasomatized by alkali-rich fluids solutions (Leroy, 1978; Racio et al., 1997).

Wind Mountain, Cornudas Mountains

About 50 mi. (80 km) east of El Paso and just north of the New Mexico–Texas boundary is the Wind Mountain, which is located in Otero County, New Mexico, and is one of the largest uplifted areas of the Cornudas Mountains. Wind Mountain was formed by a laccolith (i.e., igneous, lens-shaped rock intruded between rock strata causing uplift in the shape of a dome) of porphyritic nepheline syenite that rises 2,500 ft (762 m) above the surrounding Diablo Plateau (Holser, 1959). The main mass of the laccolith is cut by dikes and sills of nepheline syenite and syenite. Some of these sills and dikes contain thorium, uranium, and REE (McLemore, 1983b). REE deposits are typically associated with alkaline rocks, widespread alkali metasomatism, and abundant breccia (Singer, 1986). The intrusions in the Cornudas Mountains have rare-earth enrichment, but only minor altered and brecciated rock (Long and Van Gosen, 2010). The most altered area in the Cornudas Mountains is in the upper Chess Draw, which has been drilled, sampled, and evaluated as having no potential (Schreiner, 1994). Currently (2021), the potential for undiscovered REE and niobium deposits is low to moderate.

New York

Mineville Iron District

Thorium and REE reside within apatite in iron ores once mined in the Mineville, New York, area. The Mineville iron district includes iron ores once mined in the area, located in the northeastern part of the Adirondack Mountains, on the westside of Lake Champlain. Most of the former iron mines are in the towns of Mineville and Port Henty in Essex County, New York. The primary apatite-rich iron deposits are the Old Bed, Cheever, and Smith bodies; the Cheever and Smith orebodies have been mined out. The orebodies are magnetite deposits that are intricately folded and faulted within a complex suite of Precambrian metamorphic and igneous rocks. Mafic and felsite are included in the host rocks along with augite syenites, granite, gabbro, and diorite (Kemp, 1908; Staaz et al., 1980). The iron deposits are mainly magnetite, martite, and apatite, which gangue minerals of hornblende, augite, quartz, albite, pyrite, and tourmaline (McKnown and Klemic, 1956). In the Mineville-Port Henty area, the iron deposits that are high in apatite are also enriched in phosphorous, thorium, and REE because these elements are concentrated within the apatite grains. Also, the ore mineral magnetite is intergrown with 0.04–0.12 in. (1–3 mm) long, rice-shaped grains of apatite. Currently (2021), there is no apparent exploration or production in this district (Long and Van Gosen, 2010).

Wyoming

Bear Lodge Mountains

REEs–thorium deposits are exposed in the southern Bear Lodge Mountains, about 5 mi. (8 km) northwest of Sundance, Crook County, Wyoming. REE and thorium are found in veins and disseminated deposits. The later type of deposit is somewhat analogous to copper porphyries. The deposits represent a large low-grad source of these elements. REE and thorium deposits occur principally in the central part of the main intrusive mass. Most are found between Whitelaw Creek on the north and the U.S Forest Service lookout on Warren Peaks to the south. The favorable area is marked by high radioactivity (Staaz, 1983). Breccia bodies are associated with the igneous intrusions, such as a heterolithic diatreme (i.e., a long vertical pipe or plug formed when gas-filled magma forced its way up through overlying strata) breccia near Bull Hill; REE-bearing carbonatite dikes intruded near the Bull Hill diatreme; the dikes are surrounded by a large zone of low-grade REE mineralization that fills thin, narrow stockwork fractures within the large alkaline intrusions. These thorium and REE deposits crop out throughout the area of about 6 mi.^{2} (16 m^2) (Long and Van Gosen, 2010; Staaz, 1983). No mineral resources have been produced from these veins thus far (2021).

NOTE

1 Based on the U.S. Geological Survey Scientifics Report 2010-5220 compiled by Long, K.R., Van Gosen, B.S., Foley, N.K., and Cordier, D. 2010. Accessed 11/23/2021 @ http://pubs.usgsgov/sir/2010/5220/.

REFERENCES

Allen, M.S. and Foord, E.E. (1991). Geological, geochemical and isotopic characteristics of the Lincoln County porphyry belt, New Mexico: implications for regional tectonics and mineral deposits. Socorro, NM: New Mexico Geological Society, Guidebook 42, pp. 97–113.

Allen, M.S. and McLemore, V.T. (1991). The geology and petrogenesis of the Capitan pluton, New Mexico. Socorro, NM: New Mexico Geological Society, Guidebook 42, pp. 115–127.

Armbrustmacher, T.J. (1988). Geology and resources of thorium and associated elements in the Wet Mountains area, Fremont and Custer Counties, Colorado. U.S. Geological Survey Professional Paper 1049-F, p. 34, 1 plate.

Bell, K. ed. (1989). *Carbonatites: Genesis and Evolution*. London: Unwin Hyman.

Bingler, E.C. (1968). Geology and minerals resources of Rio Arriba County, New Mexico. New Mexico Bureau of Mines and Mineral Resources Bulletin 91, p. 158, 8 plates.

Brown, J.S., Emery, J.A. and Meyer, P.A., Jr. (1954). Explosion test pipe in test well on Hicks Dome, Hardin County, Illinois. *Economic Geology* 49(8): 891–902.

California Division of Mines (2007). Geologic History Special Report 68. Sacramento, California.

Castor, S.B. (2008). The Mountain Pass rare-earth carbonatite and associated ultrapotassic rocks, California. *The Canadian Mineralogist* 46(4): 779–806.

Castor, S.B. and Nason, G.W. (2004). Mountain Pass rare earth deposit, California. In S.B. Castor, K.G. Papke and R.O. Meeuwig (eds.) *Betting on Industrial Minerals—Proceedings of the 39th Forum on the Geology of Industrial Minerals*, Reno/Sparks, Nev., May 18–24, 2003. Nevada Bureau of Mines and Geology Special Publication 33, pp. 68–81.

Christman, R.A., Brock, M.R., Pearson, R.C. and Singewald, Q.D. (1959). Geology and thorium deposits of the Wet Mountains, Colorado—A progress report. U.S. Geological Survey Bulletin 1072-H, p. 515.

Christman, R.A., Heyman, A.M., Dellwig, L.F. and Gott, G.B. (1953). Thorium Investigations 1950–52. Wet Mountain, Colorado. U.S. Geological Survey Circular 290, p. 40, 5 plates.

De Saint-Andre, B., Lancelot, J.R. and Collot, B. (1983). U-Pb geochronology of the Bokan Mountain peralkaline granite, southeastern Alaska. *Canadian Journal of Earth Sciences* 20(2): 236–245.

Dunbar, N.W., Campbell, A.R. and Candela, P.A. (1996). Physical, chemical, and mineralogical evidence for magmatic fluid migration within the Capitan pluton, southeastern New Mexico. Socorro, NM: New Mexico Geological Society Bulletin 108, pp. 318–333.

Gillerman, E. (1964). Mineral deposits of western Grant County, New Mexico. New Mexico Bureau of Mines and Mineral Resources Bulletin 83, p. 213, 11 plates.

Hedlund, D.C. (1978). Geologic map of the Gold Hill quadrangle, Hidalgo and Grant Counties, New Mexico. U.S. Geological Survey Miscellaneous Field Studies Map MR-1035, scale 1:24,000.

Heylmun, E.B. (1999). Rare earths at Bokan Mountain, Alaska. *International California Mining Journal* 68(5): 44–46.

Holser, W.T. (1959). Trans-Pecos region, Texas and New Mexico. In L.A. Warner, W.T. Holzer, V.R. Wilmarth and E.N. Cameron (eds.) Occurrence of nonpegmatite beryllium in the United States. U.S. Geological Survey Professional Paper 318, pp. 130–143.

Houston, J.R., Velikanje, R.S, Bates, R.G. and Wedow, H., Jr. (1955). Reconnaissance for radioactive deposits in southeastern Alaska, 1952. Washington DC: U.S Geological Survey Trace Elements Investigations Report 293, p. 58.

Kemp, J.F. (1908). The Mineville-Port Henry Group. In D.H. Newland and J.F. Kemp (eds.) Geology of the Adirondack magnetic iron ores. New York State Museum Bulletin 119, pp. 57–88.

Keyser, H.J. and McKenney, J. (2007). Geological Report on the Bokan Mountain Property, Prince of Wales Island, Alaska: Private Report for Landmark Minerals, In., p. 48.

Lamphere, M.a., MacKevett, E.M., Jr. and Stern, T.W. (1964). Potassium-argon and lead-alpha ages of plutonic rocks, Bokan Mountain area, Alaska. *Science* 145(3633), pp. 705–707.

Leroy, J. (1978). The Margnac and Fanay uranium deposits of the La Crouzille district (western massif Central, France: geologic and fluid intrusion studies. *Economic Geology* 73: 1611–1634.

Lasley, S. (2021). Rare earths are now being mined in Canada. Accessed 12/12/21 @ https://republicofminimg.com/2021.

Long, K.R. and Van Gosen, B.S. (2010). The principal rate earth elements deposits of the united States—a summary of domestic deposits and a global perspective. Washington, DC: USGS Scientific Investigations Report 2010–5220.

MacKevett, E.M., Jr. (1963). Geology and ore deposits of the Bokan Mountain uranium-thorium area, southeastern Alaska. Washington, DC: U.S. Geological Survey Bulletin 11544, p. 125, 5 plates.

McKnown, F.A. and Klemic H. (1956). Rare earth bearing apatite. Accessed 12/20/21 @ https://mrdata.usgs.gov/deposit/show-deposit.php.

McLemore, V.T., (1983a). Carbonatites in the Lemitar and Chupadera Mountains, Socorro County, New Mexico. In C.E. Chapin (ed.) Socorro Region II, New Mexico Geological Society Thirty-fourth Annual Field Conference, October 13–15, 1983, New Mexico Geological Society, pp. 235–240.

McLemore, V.T. (1983b). Uranium and thorium occurrences in New Mexico—Distribution, Geology, production and resources, with selected bibliography. New Mexico Bureau of Mines and Mineral Resources Open-File Report 183, p. 180, text, 6 appendices.

McLemore, V.T. (1987). Geology and regional implications of carbonatites in the Lemitar Mountains, central New Mexico. *Journal of Geology* 95(2): 255–270.

McLemore, V.T. (2015). Rare earth elements (REE) deposits in New Mexico: update. *New Mexico Geology* 37(3): 59–69.

McLemore, V. (2018). Rare Earth Elements in Proterozoic Rocks. Accessed 12/15/21 @ https://geoinfor.net.edu.Staff(McLemore Projects)mining/RTE.

McLemore, V.T., Ramo, O.T., Heizler, M.T. and Heinonon, A.P. (2012). Intermittent Proterozoic plutonic magnesium and neoprotozoic cooling history in the Caballo Mountains, Sierra County, New Mexico, Preliminary Results. Socorro, NM: New Mexico Geological Society Guidebook 53, pp. 235–248.

Nuelle, L.M., Day, W.C., Sidder, G.B. and Seeger, C.M. (1992). Geology and mineral paragenesis of the Pea Ridge iron ore mine, Washington County, Missouri—Origin of the rare-earth-element- and gold-bearing breccia pipes, Chapter A. In W.C. Day and D.E. Lane (eds.) Strategic and critical minerals in the midcontinent region, United States. U.S. Geological Survey Bulletin 1989-A, pp. A1–A11.

Nuelle, L.M., Kisvarsanyi, E.B., Seeger, C.M., Day, W.C. and Sidder, G.B. (1991). Structural setting and control of the Pea Ridge magnetite deposit, Middle Proterozoic St. Francis terrane, Missouri [abs]. Geological Society of America Abstracts with Programs, vol. 23, no. 5, p. 292.

Olson, J.C., Shawe, D.R., Pray, L.C. and Sharp, W.N. (1954). Rare-earth mineral deposits of the Mountain Pass district, San Bernardino County, California. U.S. Geological Survey Professional Paper 261, p. 35.

Olson, J.C., Marvin, R.F., Parker, r.L. and Mehnert, H.H. (1977). Age and tectonic setting of lower Paleozoic alkalic and mafic rocks, carbonatites, and thorium veins in south-central Colorado. *U.S. Geological Survey Journal of Research* 5(6): 673–687.

Philpotts, J.A., Taylor, C.D., Tatsumoto, M. and Belkin, H.E. (1998). Petrogenesis of Late-stage granites and Y-REE-ZR-NB-enriched vein dikes of the Bokan Mountain stock, Prince of Wales Island, Southeastern Alaska. Washington, DC: U.S. Dept. of Interior.

Racio, C., Fallick, A.E., Ugidos, J.M. and Stephens, W.E. (1997). Characterization of multiple fluid gravata interaction processes in episyenites Avtia-bojar, central Iberian massif, Spain. *Chemical* 143: 127–144. Geology.

Richter, D.H., Sharp, W.N., Watts, K.C., Raines, G.L., Houser, B.B. and Klein, D.P. (1986). Maps showing mineral resource assessment of the Silver City 1° × 2° quadrangle, New Mexico and Arizona. U.S. Geological Survey Miscellaneous Investigations Series Map 1–1310-F, p. 24, pamphlet 1 sheet various scales.

Robertson, J.M. (1976). Mining deposits of northeastern New Mexico. In R.C. Ewing and B.S. Kues (eds.) Guidebook Vermejo Park, northeastern New Mexico. New Mexico Geological Society Twenty-Seventh Field Conference, September 30, October 1 and 2, 1976, New Mexico Geological Society, pp. 257–262.

Schreiner R.A. (1994). Mineral investigation of Wind Mountain and the Chess Draw area, Cornudas Mountains, Otero County. New Mexico: U.S. Bureau of Mines MLA 26095, p. 46.

Seeger, J. et al. (2001). Compounds of rare earth metals. Accessed 12/12/21 @ https://www.semanticssociology.org/paper.

Shawe, D.R. (1953). Thorium resources of the Mountain Pass district, San Bernardino County, California. U.S. Geological Survey Trace Elements Investigation Report 251, p. 73, 4 plates.

Sidder, G.B., Day, W.C., Nuelle, L.M., Seeger, C.M. and Kisvarsanyi, E.B. (1993). Mineralogic and fluid-inclusion studies of the Pea Ridge iron-rare-earth deposit, southeast Missouri, Chapter U. In R.W. Scott, Jr., P.S. Detra and B.R. Berger (eds.) Advances related to United States and international mineral resources—developing frameworks and exploration technologies. U.S. Geological Survey Bulletin 2039, pp. 205–216.

Singer, D.A. (1986). Descriptive model of carbonatite deposits. In D.P. Cox and D.A. Singer (eds.) Mineral deposit models. U.S. Geological Survey Bulletin 1693, pp. 51–53.

Staaz, M.H. (1972a). Thorium-rich veins of Hall Mountain in northernmost Idaho. *Economic Geology* 67(2): 240–248.

Staaz, M.H. (1972b). Geology and description of the thorium-bearing veins, Lemhi Pass quadrangle, Idaho and Montana. U.S Geological Survey Bulletin 1352, p. 95.

Staaz, M.H. (1974). Thorium veins in the United States. *Economic Geology* 69(4): 494–507.

Staaz, M.H. (1978). I and L uranium and thorium vein system, Bokan Mountain, southeastern Alaska. *Economic Geology* 73(4): 512–523.

Staaz, M.H. (1983). *Geology and Description of Thorium and Rare-Earth Deposits the Southern Bear Lodge Mountains, Northeastern Wyoming*. Washington, DC: U.S. Department of Interior.

Staaz, M.H. (1985). Geology and description of the thorium and rare-earth veins in the Laughlin Peak area, Colfax County, New Mexico. U.S. Geological Survey Professional Paper 1049-E, p. 32, 1 plate, scale 1:12,000.

Staaz, M.H., Armbrustmacher, T.J., Olson, J.C., Brownfield, I.K., Brock, M.R., Lemons, J.F, Jr., Coppa, L.V. and Clingan, B.V. (1979). Principal thorium resources in the United States. U.S. Geological Survey Circular 805, p. 42.

Staaz, M.H., Hall, R.B., Macke, D.L., Armbrustmacher, T.J. and Brownfield, I.K. (1980). Thorium resources of selected regions in the United States. U.S. Geological Survey Circular 824, p. 32.

Thompson, T.B. (1988). Geology and uranium-thorium mineral deposits of the Bokan Mountain granite complex, southeastern Alaska. *Ore Geology Reviews* 3: 193–210.

US Department of Interior (1989). *Report by J. D. Warner, Columbium, Rare-Earth-Element-, and Thorium-Bearing Veins Near Salmon Bay, Southeastern, Alaska*. Washington, DC: United States Department of Interior.

USGS (2010). Mineral commodity summaries, 2010. Accessed 11/21/2021 @ http://minerals.usgs.gov/minerals/index.html.

USGS (2017a). Rare earth elements. Professional paper 1802-0 by Van Gosen, B.S. et al and edited by Schultz, K.J. et al. Accessed 11/22/2021 @ https://doi.org/10.3133.

USGS (2017b). Alaska has considerable potential for undiscovered mineral resources including critical minerals. Accessed 11/28/2021 @ https://www.usgs.gov/new/stude-evaluates-critical-mineral-resoures-potentail-Alaska.

Van Allen, B.R., Emmons, D.L. and Paster, T.P. (1986). Carbonatite dikes of the Chupadera Mountains, Socorro County, New Mexico. *New Mexico Geology* 8(2): 25–29.

Van Gosen, B.S. and Lowers, H.A. (2007). Iron Hill (Powderhorn) carbonatite complex, Gunnison, CO—a potential source of several uncommon mineral resources. *Mining Engineering* 59(10): 6–62.

Van Gosen, B.S., Gillerman, V.S. and Armbrustmacher, T.J. (2009). Thorium deposits in the United States—energy resources for the future? U.S Geological Survey Circular 1336, p. 21. Accessed 11/22/2021 @ http://pubs.usgs.gov/circ/1336.

Vierrether, C and Cornell, W.J. (1993). *Geology, Material Science. Rare Earth occurrences in the Pea Ridge Tailings*. Atlanta, Georgia: Center for Disease Control and prevention.

Warner, J.D. and Barker, J.C. (1989). Columbium- and rare-earth element-bearing deposits at Bokan Mountain, southeast Alaska. Washington, DC: U.S. Bureau of Mines Open File Report 33–89, p. 196.

Warner J. and W. Barkley (1963). Metals and compounds of rate earths. Accessed 12/14/21 @ https://www.semanticschloro.org/paper.

Xu, A. (1996). Minerology, petrology, geochemistry and origin of the Elk Creek Carbonatite. PhD dissertation, University of Nebraska, Lincoln, p. 299.

Zharikov, V.A., Pertsev, N.N., Pertsev, V.I., Rusinov, V.L., Callegari, E. and Fettes, D.J. (2007). Metasomatism and metasomatic rocks. Accessed 12/1/2021 @ www.bgs.ac.uk.scrm/home.html.

7 U.S. Phosphorite Deposits

One thing seems certain (is certain) the critical role of rare earth elements (REE) has created a demand that is quickly outstripping known global supply and has triggered, especially in the United States, a scramble to discover, mine, and process both LREE (light) and HREE (heavy). With regard, and the author's exploration and research, for this pressing need for REE (of any type), the good news is that the chemical analysis of several sedimentary phosphate deposits (phosphorites) in the United States demonstrates that they are significantly enriched in REEs. Another good point about the presence of REE in phosphorite deposits is that they are not subject to many technological and environmental challenges (based on experiments using dilute sulfuric and hydrochloric acid) that trouble the exploitation of many identified REE deposits. A couple of interesting points about prospecting and processing phosphorite deposits of REE include, first, that phosphate rock in the United States has the potential to produce a significant proportion of the world's REE demand as a byproduct. Second, the size and concentration of REEs in some unmined phosphorites dwarf the world's richest REE deposits. The ease of REE extraction from phosphorite deposits indicates that phosphorites might be a primary source of REEs with the potential to resolve our increasing need for REE, especially HREE.

PHOSPHORITE AND REE

Phosphorite, rock phosphate or phosphate rock, is a non-detrital sedimentary rock that contains high amounts of phosphate minerals; they form as chemical precipitates on continental shelves; upwelling of cold phosphate-rich waters causes warming decreases in solubility. RREs are being found within phosphorites. This is significant because with increasing demand from modern technology another source of REE is important. The good news is that in the United States phosphorites not only afford us of another source of REE but also help to decrease our dependence on other countries (e.g., like China) to provide us with the increasing need for REE.

Phosphorite deposits are found in the southeastern United States along the Atlantic Coastal plain from North Carolina to the center of the Florida peninsula (phosphate deposits are also found in Virginia and Tennessee), forming a larger phosphogenic province that has been subdivided into the Carolina Phosphogenic Province and the Florida Province (Riggs, 1984).

With regard to Florida, three main types of phosphatic rock have been identified: land-pebble phosphate, hard-rock phosphate, and river-pebble phosphate (Cathcart, 1949). It is the land-pebble deposits that are of interest to us because they not only contain uranium but also rare-earth elements. At the present time, the most active and productive areas of the land-pebble district are in Hillsborough and Polk

DOI: 10.1201/9781003350811-8

counties, located in the west-central part of the Florida peninsula and include the upper Tertiary Hawthorn Groups and Bone Valley Formation. The Hawthorn Group is part of the Coosawhatchie Formation. The Coosawhatchie Formation is exposed or lies beneath a thin overburden on the eastern flank of the Ocala Platform from southern Columbia County to southern Marion county. Within the outcrop region, the Coosawhatchie Formation varies from a light gray to olive gray, poorly consolidate, variably clayey and phosphatic sand with few fossils, to an olive gray, poorly to moderately consolidate, slightly sandy, silty clay with few to no fossils—permeability of the sediments is generally low, forming part of the intermediate confining unit/aquifer system. Occasionally the sands will contain a dolomitic component and,

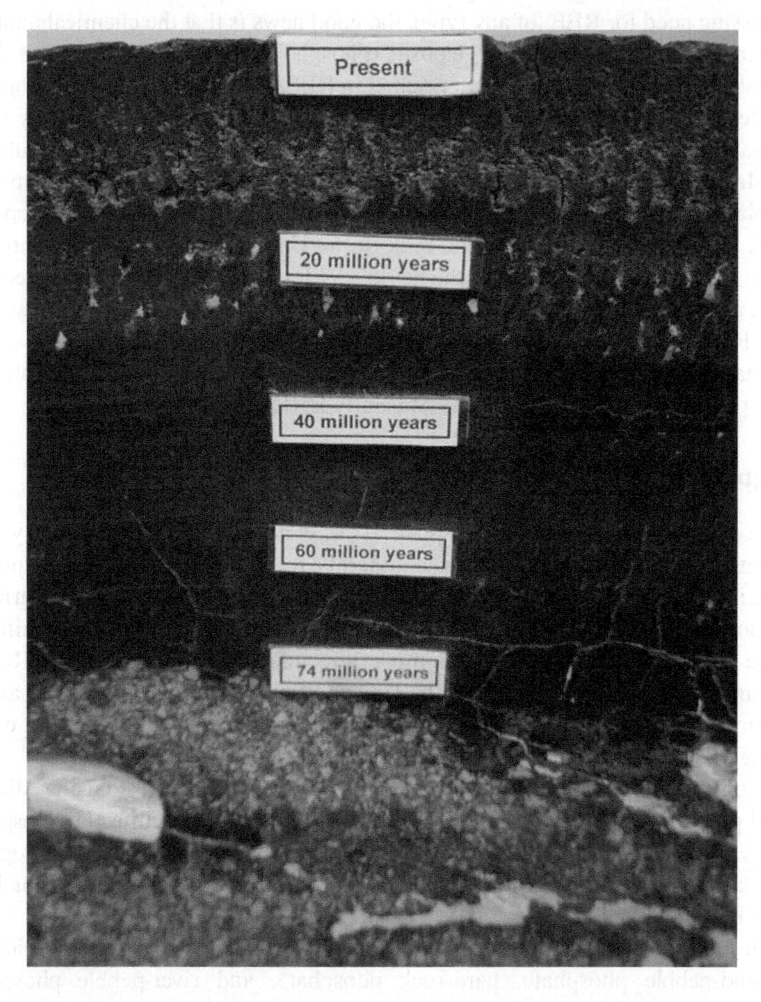

FIGURE 7.1 Cross-section of one of the thickest ferromanganese seafloor crests ever recovered. USGS (2021) public domain photo. Accessed @ https://www.usgs.gov/media/images/seafloor-crust-marshall-islands.

rarely, the dominant lithology will be dolostone or limestone Silicified nodules are often present in the Coosawhatchie Formation sediments in the outcrop region. The sediment may contain 20% or more phosphate (Scott, 1988). Mining of phosphate in this area of Florida has been concentrated in this area but there are other areas, too (Cathcart et al., 1952). This Florida region has been referred to more recently as the central Florida phosphate district (Kauwenberg and McClellan, 1999).

Overlaid by a zone of loose quartz sands and surface soil, the Bone Valley formation, a clayey and sandy phosphorite, overlies the Hawthorn in west-central Florida. The open pit mines of this deposit provide about three-fourths of the phosphate mined in the United States. Although the bone Valley formation is fossiliferous, containing shark's teeth and manatee bones in large numbers, few diagnostic fossils have been found, and the origin and age of the formation are considered open questions (Cathcart et al., 1952). Students of the problem have proposed two general assumptions: that the Bone Valley formation is a residuum derived in place by leaching the carbonates from the Hawthorn formation, and that the Bone Valle is a Pliocene deposit containing reworked Hawthorn residuum.

A geological timeline of the cross-section formation of phosphorite deposits is shown in Figure 7.1 and an example of phosphorite is shown in Figure 7.2 (both are public domain photos in USGS photo library).

FIGURE 7.2 An example of phosphorite; cross-section of barite and silica chimney. The center surrounds a conduit through which hot fluids are vented. USGS (2021) public domain photo. Accessed @ https://usgs/centers/pacific-coastal-and-marine-science/center.science.

ATLANTIC COASTAL PLAIN PHOSPHORITE PROSPECT

Phosphorite deposits are located in the southeastern United States along the Atlantic Coastal Plain from North Carolina to the center of the Florida peninsula, forming a large phosphogenic province that has been subdivided into the Carolina Phosphogenic Province and the Florida Phosphogenic Province (Riggs, 1984). Phosphate deposits are also located in Tennessee and Virginia. Florida Institute of Phosphate Research (FIPR, 2010) points out that mining for phosphate (to be used for fertilizer) in Florida dates back to 1883 in hard-rock deposits located near Hawthorn in Alachua County. Florida and North Carolina account for the bulk of production, and the majority of phosphate comes from Florida.

The bottom line: increased interest and ongoing research on the REE and trace metal contents of Florida phosphorites is critical to assess its viability more fully as an economic resource. This also applies to continuing work on the North Carolina phosphorites for assessing REE and trace metal content.

REFERENCES

Cathcart, J.B. (1949). Distribution of uranium in the Florida phosphate field. U.S. Geological Survey Trace Elements Investigation Report 85, p. 18.

Cathcart, J.B., Blade, L.V., Davidson, D.F. and Ketner, K.G. (1952). The geology of the Florida land-pebble phosphate deposits. U.S. Geological Survey Trace Elements Investigation Report 265, p. 21.

Florida Institute of Phosphate Research (FIPR) (2010). Accessed 12/17/2021 @ http://www.fipr.state.fl.us/research-area-mining-htm#history.

Riggs, S.R. (1984). Paleoceanographic model of Neogene phosphorite deposition, U.S. Atlantic Continental Margin. *Science* 223(4632): 123–131.

S.J. Kauwenberg and G.H. McClellan (1999). Minerology and alteration of the phosphate deposits of Florida. Accessed 12/12/21 @link.library.in.gov/portal/minerology-and-alteration-of-the.phophate/Smhy5x92tim/

Scott, T.M. (1988). The lithostratigraphy of the Hawthorn Group (Miocene) of Florida. Florida Geological Survey Bulletin 59, p. 148.

8 Placer Rare Earth Elements Deposits

It turns out that ancient and modern sedimentary placer deposits that were formed in coastal and alluvial deposits (a mix of silt, sand, clay, and gravel with large amounts or organic matter deposited by rivers) are and have been significant sources of rare earth elements (REEs) (Figure 8.1). Globally and in the United States, alluvial accumulations of monazite are a valuable type of REEs–thorium (Th) deposit. Monazite (see Figure 8.2), a high-density (heavy) mineral, is the primary source of placer REE-bearing minerals. One of the World's most significant accumulations of monazite containing thorium is located along the Coastal Belt of southernmost India. This particular deposit contains detrital heavy minerals and is found in piedmont lakes, shallow seas, parts of beaches, sand bars across the of rivers, deltas, and sand dunes behind the beaches (Bhola et al., 1958). A 1958 study by Mahadevan et al. (1958) estimated that the beach sands of the southwestern coast of India alone contain estimated reserves of more than 440,000 metric tons (492,000 tones) of monazite, in which the ThO_2 content of the monazite ranges from 7.5 to 9%.

Because of their high density and high specific gravity monazite and its resistance to chemical weathering account for its association in place (alluvial) deposits with other resistant heavy minerals such as ilmenite, magnetite, rutile, and zircon. Monazite weathers from alkaline crystalline rocks of the surround region and are transported downstream and deposited by alluvial processes. Monazite has been extracted from many heavy mineral placers as a coproduct of the economic recovery of associated industrial minerals such as titanium oxide minerals, and others (Long and Van Goshen, 2010; Sengupta and Van Gosen, 2016).

In the United States, placer (alluvial) deposits of monazite are known in the Carolina Piedmont of North and South Carolina, the beach deposits of northeastern Florida through southeastern Georgia, and the intermontane valleys of Idaho. Monazite REEs are also located placer, alluvial deposits in Alaska, Wyoming, Virginia, and Tennessee (see Table 8.1).

One of the primary REE in monazite is monazite–(Ce)—Cerium (Ce). Monazite also contains lesser amounts of La, Nd, Sm, and Pr. Monazite is also an important ore for thorium but Ce is the most common member (see Figure 8.2).

With regard to unconsolidated stream deposits in Idaho and North and South Carolina, in the past, these were mined by small-scale sluices and a few in Idaho by dredges; beach sands in Florida and Georgia were mined by large shovels. Sluicing is often the choice by small-scale operations because it instantly produces a heavy mined separate, however, even in the case of the dredge or shovel operations the heavy-mineral separation was still performed at the site. The huge benefit of mining REE and thorium from alluvial deposits is the advantages of the relative ease of mining and rapid mineral separation, in contrast to hard-rock mining. Another benefit of

DOI: 10.1201/9781003350811-9

FIGURE 8.1 Monazite.

Source: Public domain photo accessed 12.16/2021 @ https://www.mindat.org/photo.

FIGURE 8.2 Monazite containing Ce. Photo by F. Spellman (2022).

place deposits is the potential for coproducts. Coproducts can be important and valuable because they can include REE and thorium obtained from monazite; titanium from ilmenite and rutile; iron from magnetite; zirconium and hafnium from zircon; and industrial-grade garnet, staurolite, tourmaline, kyanite, and sillimanite, which are used as abrasives and refractory minerals (have high resistance to temperature) (Long and Van Goshen, 2010).

Now let's take a look at the largest (to date, 2021) volume alluvial REE–thorium deposits known in the United States: North and South Carolina stream deposits,

TABLE 8.1
United States Monazite REEs Deposits

Name of Deposit	Type Deposit	State
McGrath	Placer, alluvial	Alaska
Port Clarence	Placer, alluvial	Alaska
Tolovana	Placer, alluvial	Alaska
Pearson Creek	Placer, alluvial	Idaho
Bear Valley County	Placer, alluvial	Idaho/valley
Big Creek County	Placer, alluvial	Idaho/valley
Gold Fork-Little Valley County	Placer, alluvial	Idaho/valley
Hilton Head Island	Placer, alluvial	South Carolina
Horse Creek	Placer, alluvial	South Carolina
Green Cove Springs	Placer, alluvial	Florida
Maxville	Placer, alluvial	Florida
Trail Ridge	Placer, alluvial	Florida
Brunswick-Altamaha	Placer, alluvial	Georgia
Cumberland Island	Placer, alluvial	Georgia
Natchez Trace deposit	Placer, shoreline	Tennessee
North Camden	Placer, shoreline	Tennessee
Oak Grove	Placer, alluvial	Tennessee
Oak Grove	Placer, paleoplacer	Tennessee
Silica Mine	Placer, alluvial	Tennessee
Old Hickory	Placer, shoreline	Virginia
Bald Mountain	Placer, paleoplacer	Wyoming

Source: F. Spellman (2009). *Geology for Nongeologists*. Lanham, MD: Scarecrow Press.

Florida–Georgia beaches. The geology and estimated monazite resources of these districts have been initially well described by Staaz et al. (1979), the North and South Caroline placer deposits, pp. 9–18, and the Florida beach deposits, pp. 3–9); numerous references cited therein and verified by several other later researchers provide not only more information but also confirmation of the researchers' findings. Herein, we only briefly touch on their findings of earlier studies that have subsequently been substantiated more than once.

PLACER DEPOSITS—IDAHO

There are at least 11 monazite-bearing placer districts in the valleys extending north of Boise, Idaho, and along the length of the western flank of the Idaho in a very large igneous intrusion extending deep in the earth's crust (i.e., a batholith; a type of igneous rock that forms when magma rises into the earth's crust but does not erupt onto the surface). Within this region, monazite-bearing alluvial stream deposits (placers) are extensive. The Idaho batholith is thought to be the primary source of resistant REEs—thorium-bearing minerals, especially the quartz monazite and pegmatite phases of the batholith (Mackin and Schmidt, 1957). Listed as follows in generally decreasing amounts, the most common heavy minerals in the alluvial deposits in the Idaho batholith are ilmenite, magnetite, sphene (aka titanite—a greenish-yellow mineral consisting of a silicate of calcium and titanium) garnet, monazite, euxenite, zircon, and uranothorite (uranium-rich thorite). As well as REE and thorium from monazite euxenite, the Idaho placer deposits (and the abandoned historic dredge waste piles) contain co-products of titanium (in ilmenite), and niobium and tantalum (in euxenite) (Long and Van Gosen, 2010).

As mentioned, it is the monazite-bearing material that hosts the REE minerals. Monazite was first recognized in the Idaho placers in 1896 as a heavy, yellow to brownish-yellow mineral that collected with other heavy minerals and gold within the sluice boxes of the gold placer operations in the Boise Basin near Idaho City, Centerville, and Placerville (Lindgren, 1897). In 1909, the Centerville Mining and Milling Co. built a mill designed to capture the Monazite. Only a miniscule amount of monazite concentrate was produced for its thorium content before a forest fire burned the mill down in 1910.

In the 1950s, in Long Valley and Bear Valley, in west-central Idaho, were where mined by dredges for monazite recovery. In September 1950, three former gold dredges converted to recover monazite worked Long Valley (under the sponsorship of the U.S. Atomic Energy Commission). Argal (1954) and Staaz et al. (1980) described the history of these dredging operations. The heavy minerals recovered in the Long Valley district were dominated by 84% ilmenite, followed by 8% monazite, 5% garnet, and 3% zircon. During this 5-year period, Staaz et al. (1980) estimated that the three dredges recovered more than 7,000 tons (about 6,430 metric tons) monazite containing 297 tons (269 metric tons) of thorium oxide. Note that the dredging ended here in mid-1995 when the government stockpile order was fulfilled. Also note, that more recently in 2003 Murray described, in great detail, the history of dredging of monazite in Idaho.

It is interesting to note that REEs and thorium were also unintentionally within the minerals euxenite and monazite from the Bear Valley placers. Under a Federal contract, the Bear Valley placers were worked by first one dredge in 1955, and then a second in 1956, with the intent to recover Nb and Ta. According to Staaz et al. (1980) "from alluvium of bear valley, 2,049nshort tons91,858 metric tons) of euxenite, 83.5 tons (75.7 metric tons) of columbite, and 54,863 tons (49,760 metric tons) of ilmenite were recovered." Note that no records of monazite recovery were kept.

PLACER DEPOSITS—NORTH AND SOUTH CAROLINA

Monazite-bearing alluvial stream deposits (placers) exist in North Carolina and South Carolina between the Catawba River and in the northeast and the savannah river in the southwest, along a belt that extends from east-central Virginia southwestward into Alabama. The stream-sediment deposits in this region are generally consistent; the heavy-mineral concentrations are greatest in the headwaters areas (Long and Van Goshen, 2010). It is in the flat valleys where the alluvium is deposited, forming well-bedded, poorly graded layers of unconsolidated sediment. Stacked layers, with an average total thickness of about 15 ft (4.5 m), contain gravel, sand, silt, and clayey silt (Staaz et al., 1979). Monazite is least likely to be found in clayey units but is quite abundant in basal gravel layers. Staaz et al. (1979) point out that the heavy content of the other placer deposits of the Piedmont region ranges from 0.15 to 2.0%; monazite makes up about 3.5–13% of the heavy minerals. Ilmenite, garnet, rutile, and zircon make up other parts of the mineral fraction. In some placers, additional heavy minerals like sphene, magnetite, and others (Long and Van Goshen, 2010).

Note that in 1887, a few short tons of monazite were mined from stream deposits in the Piedmont region of North and South Carolina, giving this region the distinction of being world's first supplier of thorium (Olsen and Overstreet, 1964). Small-scale sluice operations of monazite-bearing placers were worked from 1887 to 1911 and then from 1915 to 1917; they produced a 5,483 tons (4,973 metric tons) of monazite (Olsen and Overstreet, 1964). By 1917, monazite mining in this area ended because beach deposits from India and Brazil were producing thorium at lower cost.

With regard to the monazite placers in the Piedmont region of North and South Carolina, the stream-sediment deposits are consistent in character; the heavy-mineral concentrations are greatest in the headwaters areas. The flat valleys in the regions are the depositories of alluvium where well-bedded, poorly graded layers of unconsolidated sediment are formed. Stacked layers contain gravel, sand, clay, and clayey silt, at an average thickness of approximately 15 ft (4.5 m) (Staaz et al., 1979). Although monazite is usually found in all units, it is usually more abundant in basal gravel layers and least abundant in the clay layers (Long and Van Goshen, 2010).

Crystalline, high-grade metamorphic rocks intruded by quartz monzonite and pegmatite underlie the Piedmont region. The intrusions may or may not be monazite bearing. Overstreet (1967) suggested that the primary source of the alluvial monazite was the high-grade metamorphic rocks, particularly sillimanite schist. Mertie (1975) pointed out that other metamorphic rocks in this area include mica and hornblende gneisses, amphibolites, and additional varieties of schist. Several other igneous rocks are present (Long and Van Goshen, 2010).

Monazite and other heavy minerals are carried by rivers eastward from the Piedmont region, in a manner that all regions of the Coastal Plain may have also received assorted amounts of heavy minerals (Staaz et al, 1979). For instance, the Late Cretaceous Tuscaloosa Formation received a hefty amount of monazite, although widely dispersed. Even though this area has not been as well explored as the Piedmont. It is known that the Tuscaloosa directly overlaps the crystalline rocks of Piedmont and that streams in the area have modified the Tuscaloosa sand so that in places heavy-mineral placers containing monazite have been identified (Staaz et al., 1979). Horse Creek placer is best known, which is located in Aiken, South Carolina, and was the site of the first large-scale mining of stream placers for monazite and other heavy minerals in the Carolinas. In the summers of 1955 and 1958, dredging found heavy-mineral contents of about 1–5%, of which a fraction monazite formed about 8% (Mertie, 1975; Williams, 1967). On the whole, these dredging operations recovered monazite, zircon, rutile, ilmenite, and staurolite (Williams, 1967).

The sharp topographic break that marks the boundary between the Piedmont and the Coastal Plain is known as the Fall Line. East of the Fall Line, the heavy-mineral

FIGURE 8.3 Staurolite. Photo by F. Spellman.

distribution in two deposits shows several differences from that of the Piedmont. First, the abundance of staurolite increases to 7 and 38% of the abundance of the two deposits (a brown glassy mineral consisting of silicate of iron and aluminum, which occurs as hexagonal prisms often twinned with in the shape of a cross; see Figure 8.3; Kline et al., 1954; Mertie, 1975). Rutile (i.e., consisting of titanium oxide) and zircon are also more abundant in these deposits than the Piedmont placers. However, monazite concentrations are similar to those in placers in the Piedmont.

BEACH PLACER DEPOSITS—FLORIDA–GEORGIA

In placer mining when a location is found that contains a large tonnage of suitable beach deposits that are readily accessible, where mining and processing can be accomplished with relative ease and the location possesses minerals that are highly sought after this is, in this author's view, a "golden" place. The gold? Well, REEs and thorium can be classified as "golden"—especially in this area where REE are highly sought after for number of technological applications, some of which are out there but not yet identified—to be discovered, developed in the future (Long and Van Goshen, 2010).

So, what are we talking about here? We are talking about the modern (and not so modern) raised Pleistocene and Pliocene beach placer deposits of northeastern Florida and southeastern Georgia that host low-grade but persistent concentrations of monazite. While heavy minerals constitute a small part of the beach sands, and monazite forms a small part of the heavy minerals because of the large tonnage of suitable beach sand deposits they do represent a potential REE and thorium resource. Staaz et al. (1980) estimated beach placer deposits of this region contain total of 218,000 tones (198,000 metric tons) of REE oxides, 16,200 short tons (14,700 metric tons) of thorium oxide, and 1,640 tons (1,490 metric tons) of uranium oxide, all of which are hosted in 330,000 metric tons (364,000 tons) of monazite (Long and Van Goshen, 2010).

Looking back at the history of mining monazite in the Pleistocene beach deposits of Florida were once the domestic suppliers of monazite. Staaz et al. (1980, p. 3) note, "During 1978 monazite was produced from two of the three operating heavy-mineral deposits in Florida: Titanium Enterprises at Green Cove Springs and Humphrey Mining Corp. at Boulogne recovered monazite as a byproduct." Trace amounts of monazite were also mined from the large Trail Ridge orebody south of Jacksonville in 1949 by E.I. du Pont de Nemours and Company.

In general, the monazite-bearing sands in the raised Pleistocene and Pliocene beach deposits lie as much as 50 mi. (80 km) inland, making them deposits of former shorelines (pointing out that sea level rise is nothing new). These relict (i.e., a remnant from an earlier and different environment) shorelines, which lie 10–108 ft (3–33 m) above sea level, have been noted in the outer coastal plain region from Maryland to Florida. These relict shorelines were once referred to as "marine terraces," they were divided into different levels on the basis of elevation and tectonic stability and then correlated with interglacial stages (Doering, 1960; MacNeil, 1950). Further investigations uncovered that the relict shorelines more closely bear a resemblance to barrier islands, suggesting that the coastal plain was deformed during the Pleistocene (Winkler and Howard, 1977). Because monazite includes radioactive thorium, future monazite exploration in the eastern United States coastal plain can benefit from several aeroradiometric maps (i.e., maps of data generated by aerial sensing of radiation emanating from the earth's surface) that were compiled and interpreted by the USGS (Force et al., 1982; Grosz, 1983; Grosz et al., 1989; Owens et al., 1989).

From crystalline rocks of the Piedmont province (Mertie, 1953), monazite and heavy minerals in the relict shorelines deposits of southeastern United States were eroded and carried toward the Atlantic Ocean by streams and rivers, and eventually redeposited by coastal processes. Several processes are involved with the natural concentration of heavy minerals in the shoreline area that involves transport by longshore drift; gravity separation by specific gravity, particle size, and shape; differential chemical weathering (Neiheisel, 1962); wave action; and, in some parts of the coastal environment, the actions of tides. All these forces rework the sediments in the shoreline environments through time and naturally concentrate the heavy minerals (Force, 1991).

The bottom line. Despite the low concentrations of monazite (and thus, REE and thorium) in the typical coastal placer deposit of the southeastern United States, these deposits have three distinct advantages as potential sources of REE and thorium: they are relatively easy to excavate; it is relatively easy to separate the heavy-mineral fraction in the field; and they contain several marketable mineral products. Mining in both beach placers and on relict shorelines is possible using open-pit methods, and the thickness of overburden is rarely more than 15 ft (4.5 m), on average. Also, the groundwater level in Florida and Georgia is shallow enough to conduct mining of monazite and heavy minerals using floating dredges on a pond (Staaz et al., 1980).

REFERENCES

Argal, J. (1954). Rare Earths Inc. Rare Earth Correspondence. Accessed 12/13/21 @ https://-diglib.amphior.org/sheet/rareearths.

Bhola, K.L., Chatterji, B.D., Dar, K.K., Mahadevan, C., Mahadevan, V., Metha, N.R., Handi, N.R., Nanhi, H., Nanyandas, G.R., Sahasrabudhe, G.H., Shirke, V.G. and Udas, G.R. (1958). *A survey of uranium and thorium occurrences in India.* In Proceedings of the Second United Nations International Conference on the Peaceful uses of Atomic Energy, Geneva, Switzerland, September 1–13, 1958—Volume 2, *Survey if raw material resources*, Geneva United Nations Publication, pp. 100–102.

Doering, J.A. (1960). Quaternary surface formations of southern part of Atlantic Coastal Plain. *Journal of Geology* 68(2): 182–202.

Force, E.R. (1991). Geology of titanium-mineral deposits. Geological Society of America Special Paper 259, p. 112.

Force, E.R., Grosz, A.E, Loferski, P.J. and Maylon, A.H. (1982). Aeroradioactivity maps in heavy-mineral exploration-Charleston, South Carolina, area. U.S. Geological Survey Professional Paper 1218, p. 19, 2 plates.

Grosz, A.E. (1983). Application of total-count aeroradiometric maps to the exploration for heavy-mineral deposits in the Coastal Plain of Virginia. U.S. Geological Survey Professional Paper 1263, p. 20, 5 plates.

Grosz, A.E., Cathcart, J.B., Macke, D.L., Knapp, M.S., Schmidt, W. and Scott, T.M. (1989). Geologic interpretation of the gamma-ray aeroradiometric maps of central and northern Florida. U.S. Geological Survey Professional Paper 1461, p. 48, 5 plates.

Kline, S.B. et al., (1954). Rare earth is secondary waste. Accessed 12/12/21 @ https://pubmed.ncji.nih.gov.

Lindgren, W. (1897), Monazite from Idaho. *The American Journal of Science*, Fourth Series, 4: 63–64.

Long, K.R. and Van Gosen, B.S. (2010). The principal rate earth elements deposits of the United States—a summary of domestic deposits and a global perspective. USGS Scientific Investigations Report 2010–5220, Washington, DC.

Mackin, J.H. and Schmidt, D.L. (1957). Uranium and thorium-bearing minerals in placer deposits in Idaho. Idaho Bureau of Mines and Geology, Mineral Resources Report, 7, p. 9.

Mahadevan, V., Narayana Das, G.R. and Nagaraja Roa, N. (1958). *Prospecting and evaluation of beach placers along the Coastal Belt of India.* In Proceedings of the Second United Nations International Conference on the Peaceful Uses of Atomic Energy, Geneva, Switzerland, September 1–13, 1958—Volume 2, *Survey of raw material resources*, Geneva, United States Publication, pp. 103–106.

Mertie, J.B. (1953). Monazite deposits of the southeastern Atlantic States. U.S. Geological Survey Circular 237, p. 31, 1 plate.

Mertie, J.B. (1975). Monazite placers in the southeastern Atlantic States. U.S. Geological Survey Bulletin 1390, p. 41.

Murray, S.A. (2003). Dredging monazite in Idaho. Accessed 12/18/21 @ https://www.icmh.com/magazine/article.

Neiheisel, J. (1962). Heavy-mineral investigation of recent and Pleistocene sands of Lower Coastal Plain of Georgia. *Geological Society of America Bulletin* 73(3): 365–374.

Olsen, J.C. and Overstreet, W.C. (1964). Geologic distribution and resources of thorium. U.S. Geological Survey Bulletin 1204, p. 61.

Overstreet, W.C. (1967). The geologic occurrence of monazite. U.S. Geological Survey Professional Paper 530, p. 327, 2 plates.

Owens, J.P., Grosz, A.E. and Fisher, J.C. (1989). Aeroradiometric map and geologic interpretation of part of the Florence and Georgetown 1° × 2° quadrangles, South Carolina. U.S. Geological Survey Miscellaneous Investigations Series Map 1–1948-B, 1 sheet, scale 1:250,000.

Sengupta, D. and Van Gosen, B.S. (2016). Placer-type rare earth element deposits. Accessed 12/17/2021 @ https://doi.org/10.5382/Rev.18.04.

Spellman, F.R. (2002). *Science of Geology*. Boca Raton, FL: CRC Press.

Staaz, M.H., Armbrustmacher, T.J., Olson, J.C., Brownfield, I.K., Brock, M.R., Lemons, J.F, Jr., Coppa, L.V. and Clingan, B.V. (1979). Principal thorium resources in the United States. U.S. Geological Survey Circular 805, p. 42.

Staaz, M.H., Hall, R.B. Macke, D.L., Armbrustmacher, T.J. and Brownfield, I.K. (1980). Thorium resources of selected regions in the United States. U.S. Geological Survey Circular 824, p. 32.

Williams, L. (1967). Heavy minerals in South Carolina. Columbia, S.C. South Carolina State Development Board, Division of Geology, Bulletin No. 35, p. 35.

Winkler, C.D. and Howard, J.D. (1977). Correlation of tectonically deformed shorelines on the Atlantic coastal plain. *Geology* 5(2): 123–127.

Part II

Environmental Aspects

9 Life-Cycle of REE Mines

In order to gain understanding of REE mining and its associated environmental footprint, there is a need to know and understand the REE mining process. One thing seems certain; with the increasing importance of REE for the manufacture of modern devices upon which society has become accustomed to and reliant upon along with uncertain supplies, is encouraging exploration and development of new mining sites. While it is true that REE are an important resource needed to sustain out modern technologies, the waste footprint and environmental impact of REE mining operations are expected to be as significant as current mining practices for metals and minerals. The truth be told opening a new REE mine (or any other type of mine) is not typically a linear operation; instead, the requirements, regulations, and financial obligations and assurances for a new mine are complex, nonlinear, non-sequential, and take years of planning and a lot of tire-kicking by regulators. Additionally, the economic feasibility of discovered deposits must be verified, proved, and environmental effects to the local communities and habitat also must be evaluated to determine feasibility. Another bit of truth is that it may take up to ten years to perform the process of exploration, development, and construction typically required before mining can begin. Except for a few locations, known rare earth deposits are generally considered small to medium reserves. This chapter presents a discussion of the typical process steps used in developing a new REE deposit and the associated mining wastes that typically would result. It is important to keep in mind that the mining of REE is not expected to be different than other hard rock or metal mining operations. With one major exception; that is, radioactivity of uranium and thorium. Otherwise, potential REE waste emissions can be compared to a typical hard rock mine.

ACTIVE REE MINING IN THE UNITED STATES

Before presenting the life cycle of REE mining it is important to point out that much of the material presented in this book is dated. As of this writing, it is necessary to point out that experts who have studied, mined, tested, and researched possible uses REE are also, to a degree, dated: 1950s, 1960s, 1970s, 1980s, 1990s, 2000–2021; thus, much of the information contained within this book is based on this dated material. However, this is not a drawback because the dated information provides a solid foundation for the current research, technological development, and operation of mining and processing REE.

Why is the information not more current?

Good question. The answer at this moment in time (2021) is somewhat cloudy but is experiencing, that is undergoing a clearing trend—more so with the passing day. The simple matter and fact are that currently technologists are finding more and more uses of REE in our advancing technological world, and especially in the

DOI: 10.1201/9781003350811-11

western world where maintenance of the good life is now dependent, to a degree, on the further development of REE for their growing utility, benefit, efficacy, uses, and applications including but not limited to use as or in:

- metal alloys for the aerospace industry
- ceramics
- metal alloys
- lasers
- fuel efficiency
- microwave communication for satellite industries
- color television
- computer monitors
- temperature sensors
- military defense applications
- batteries, including electric vehicle (EV) batteries
- catalysts for petroleum mining
- high-tech digital cameras
- laptop batteries
- X-ray films
- polishes
- for glass
- searchlights
- fluid-fracking catalysts
- high-temperature magnets
- liquid crystal displays
- highest power magnets known
- solar panels
- glass colorant
- electric motors
- wind turbines

So, another question is with of the numerous actual REE uses listed above, what is the United States currently doing about its need for this critical resource? Answer: a mining company is determined to restore U.S. REE supply chain and in 2020 MP Materials, along with two other companies, received a U.S. Defense Department grant intended to return rare earth production to the United States. MP Materials was awarded a Defense Production Act Title III grant to begin to refine the strategic minerals at its Mountain Pass, California, mine. President Trump signed Executive Order 13817, which—among other items—seeks to return REE production to the U.S and break China's dominance over the supply chain. Currently, the Mountain Pass mine is the largest producer of rare earth elements in the Western Hemisphere and constitutes 15% of the world's production of the minerals, which are essential to manufacture higher performance magnets. China currently has a near monopoly on refining the elements and producing the magnets, which are used in a variety of weapons systems as well as commercial goods such as wind turbines and EVs.

Gas Station or Battery Station

With regard to EVs you may recall that the rechargeable batteries that replace hydrocarbon fuels in EVs (hybrid automobiles) are constructed of lithium-ion; the substance that takes and stores an electrical charge to power the car. You may recall from our earlier discussion that lithium although it is a critical material is not an REE. So, you may be wondering if lithium is the preferred mineral for EV batteries (and other batteries) what does REE have to do with the EV battery? Well, quite a lot, actually. Inside the nickel metal hydride (NiMH) rechargeable battery, the negative electrode of the battery is a mixture of metal hydrides—one of which is typically lanthanum (an REE). Moreover, REE is used in the EV motor. The lithium-ion battery provides power to the vehicle's electric motor—the electric motor contains REE magnets, very powerful permanent magnets. These magnets provide the magnetic field that the motor rotor cuts as it spins and in turn produces an electrical current that powers the motor enabling the vehicle to move; making it drivable. Of course, the power provided by the lithium-ion battery is limited to providing power to the motor in the approximate range of 300-mile driving range only. When the battery reaches the point where it must be recharged, it must be recharged at home or at one of the few charging stations that are spread out here there and someday everywhere. Because of the short range of the EVs and because they must be recharged relatively frequently and while charging the EV is out of commission, basically, until recharged this is or can be a true inconvenience, for sure. Thus, while EVs are attractive for the so-called "greenies" because they do not pollute and help in combating climate change they do have their limitations. However, note that based on personal experience and non-scientific information provided to me by several different EV owners their love for their electrical vehicle climate-saving was quickly discharged, so to speak, and they eventually traded the EV in for another hydrocarbon-powered vehicle.

Why? There were assorted reasons but the one most frequently mentioned to me was the need to recharge the vehicles, the lack of charging stations in certain places, and also the time required to charge up the EV.

Now it is true that there is an effort underway to install thousands of additional charging stations throughout the United States but there is still the problem with the time it takes to recharge the lithium-ion vehicle battery.

The good news? There is a relatively simple solution to solve this inconvenience. When our traditional vehicles need fuel we pull into a gas station and fuel up, gas up, and so forth. Why can't we pull our EV into a battery station (versus a gas station) and change out the battery when needed with a freshly charged battery? Paying a small fee for this service probably could be less costly than the cost of a fill-up for most standard gasoline- or diesel-powered vehicles and you simply leave the discharged battery at the station where it can be recharged and placed on the shelf to be changed out for the next vehicle requiring a fresh battery.

So, is this modest proposal realistic, possible, doable, achievable, or more importantly is it feasible?

Yes, change-out-batteries are possible with a few adjustments. That is, it is possible with a few technological changes or advances. First, every exchangeable battery must be the same—the same size, capacity, and designed to easy fit into any automobile or truck—for standard vehicles and the same for trucks; they must be of universal design to power any make of model of vehicle. Second, the batteries must be easy to remove and easy to install. Again, technological changes to current EV batteries must be made because currently they are bulky and heavy and also dangerous if not handled correctly and safely.

One last requirement is necessary: the change out from discharged to fully charged EV battery must be quickly effected, 24–7, the change out taking only about the same amount of time required to gas up a standard vehicle, or fairly close to it.

The bottom line: if we are going to make a wholesale switch-over from gas/diesel-powered vehicles then we need to do some forward-thinking about how we can make changing out and EV battery at a battery station as easy a filling up at a gas station.

RARE EARTH ELEMENT ORES

Concentrations of rare earth oxides (REO) minerals (i.e., metals are the elemental constituent of REO ores) typically occur in igneous and metamorphic rocks and also in sedimentary placer deposits. Concentrations of the REO in the ore have implications for the selection of the mining method needed to obtain adequate quantities of the ore; the processing steps required and the volumes of wastes that must be managed (USEPA, 2012).

Earlier in this book we discuss the location of both hard rock and placer deposits in the conterminous United States and Alaska, along with the more common REE minerals (monazite and others). The REE-containing minerals in a hard rock or placer deposit may or may not be viable for mining due to assorted reasons, including economic, technical, and environmental reasons, stemming from the concentration of the REE-bearing minerals in the rock. The definition of an "ore" for the purposes of this book is as follows:

> … a mineral source from which a valuable commodity (e.g., metals) is recovered.
>
> The term ore implies economic viability, given the concentration of metal in the host rock, the costs of extraction, processing, and refinement, waste management, site restoration, and market value of the metal.
>
> *(USEPA, 2008)*

As the definition implies, the potential environmental impacts from all states of a mine must be considered in evaluating the feasibility of the mine start-up—including the costs of mine restoration—to determine if a REE deposit represents an ore from which REOs and REMs can be recovered economically.

MINE LIFE-CYCLE

Four life-cycle stages are recognized in this book. These life-cycle stages include exploration, mine development, ore extraction and processing, and mine closure (see Figure 9.1). Keep in mind that there are additional steps or phases incorporated within each stage. Keep in mind that additional steps or phases are also incorporated

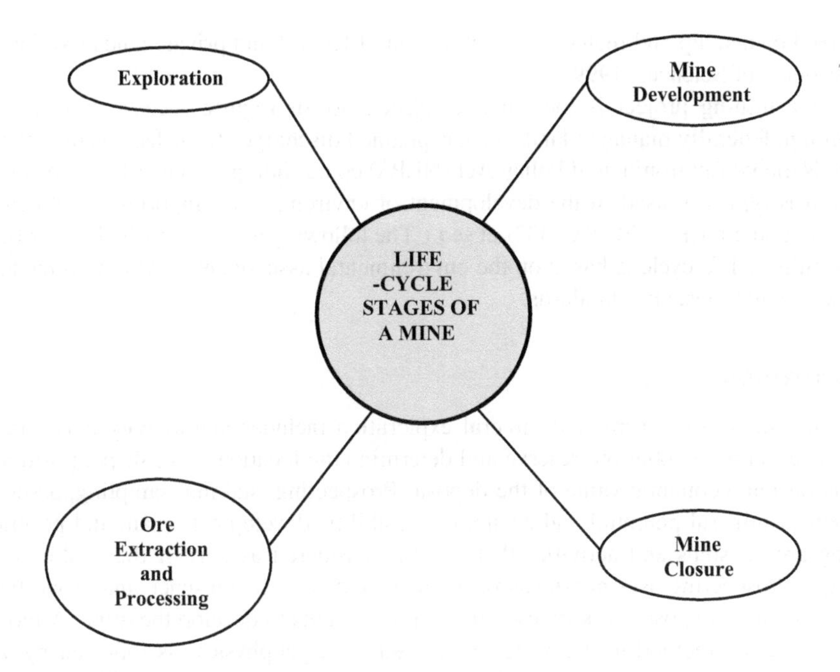

FIGURE 9.1 Four life-cycle stages of a mine.

within each stage. The activities associated with each stage can be considered as generally linear, but overlap is common. Although a linear process, in many respects, this does not mean the process is timely or quick; quite the opposite, the time needed to complete each of the steps for mine startup depends on many factors and can potentially require many years before the mine is in full production.

REE Permitting Requirements

A detailed discussion of applicable mining regulatory policy and permitting requirements developed to manage the environmental impacts from mining operations is beyond the scope of this document. For those seeking a detailed summary of applicable federal regulations they are referred to *EPA and Hardrock Mining: A Source Book for Industry in the Northwest and Alaska* (January 2003). Note that projects with potential to have significant environmental impact are regulated under federal or state permitting programs. Mines developed near tribal lands require consultation with tribal leadership, especially if the mining activity or associated operations could impact the community or subsistence resources. It is important to point out that mining operations on the scale that would be expected for the recovery of REE minerals would potentially require large areas to accommodate an individual mine and its associated support facilities (e.g., waste material management areas and tailings ponds; dumping grounds after the good stuff has been removed). This need for large areas of land typically means that the land occupied by a min is of mixed ownership and includes a combination of federal and private lands (National Academy of Sciences, 1999).

If a mining project is located on, adjacent to, or requires access and egress through federally managed lands, or has planned discharges to surface waters, then the National Environmental Policy Act (NEPA) comes into play. The life-cycle paradigm is typically used in the development of environmental important statements (EIS) (under NEPA 42 U.S.C. 4321 et seq.). The following discussion which describes the mining life-cycle is based on the environmental assessment (EA) requirements, laws, regulations, and standards.

Exploration

In the simplest of terms, the mineral exploration includes any activity performed to discover a potential ore reserve and determine the location, size, shape, position, and current economic value of the deposit. Prospecting, staking, sampling, assessment of mineral potential and economic feasibility, development plan, and permitting are all steps and activities that can be considered as part of the exploration stage. Prospecting uses non-invasive methods to assess the presence and properties of a potential deposit. Staking establishes miner rights to develop the mine. A more sophisticated method to delineate the deposit (e.g., geophysics), is followed by an intermediate stage of exploration that uses trenching, core or rotary drilling, bulk is a sample (invasive methods) to sample the deposit. Samples are examined for metal content, and the data collected from the previous steps are used to create mineral prospective maps showing the geologic favorability of the deposit; definition drilling is commonly performed during this progressed stage of exploration. If the analysis proves the deposit economically viable, then a mining development plan is drafted and applicable permits are obtained. The use of invasive assessment methods can be considered as an independent stage in one of the steps. Baseline environmental studies may be conducted, if required, prior to exploration to determine the presence of sensitive species and habitat that might be impacted during these activities (Spellman, 2017; USEPA, 2012).

The approach and methodology used for exploration depend upon various factors, such as terrain, nature, and size of the target, and expected depth of the ore body; regulatory and permitting requirements and restrictions; available information and technology; and available capital resources. Proof of the deposit for sale, and mine developing by the exploration entity is included in the scope of the initial exploration project and the approach and methods used to prove the deposit. The methods used may be simple and low cost, or technically sophisticated. The extent and type of exploration activities that can occur are influenced by the location of the mine. Note that if the deposit is located in an existing mining district and/or on privately owned lands, the drilling permits may be more easily obtained by the exploration or mining company, and more aggressive exploration such as exploratory drilling may occur. However, if the site of the deposit is located near national forest lands, then exploration activities requiring land disturbance may not be allowed until a preliminary environmental informant document (EID) is completed and undergoes EPA/NEPA review. Upon approval by the EPA or state agencies, drilling data can be collected and analyzed. The exploration data should then be used to inform additional EA that may be needed in the mine development-construction stage. Even though many states require permits sometimes environmental impacts of exploration activities are overlooked. With regard to locations where permits are required Alaska is a good example; there permits required for exploration activities include a Clean Water Act 404 permit for wetlands disturbance, camp permits; a temporary water use permit; and overland travel permit; land-use permits for off-claim camps or staging areas; hare rock exploration land-use permit; and a state bond pool requirement for reclamation. Drilling and trenching can be extensive, long-term operations using multiple types of equipment with associated waste (e.g., equipment fluids, lubricants and greases, solvents, and other traditional industrial materials). Additionally, mining-type wastes (e.g., wastewater and waste rock) can be produced during exploration activities and will need to be managed (USEPA, 2012).

Passive Mine Exploration Methods

In REE mining, a combination of traditional prospecting approaches using water, rock, soil, and sediment sampling along with shallow pits or with geologic mapping and general field investigations used in conjunction with sophisticated techniques employing geophysics, aerial and satellite remote sensing, mineral deposit models, and potentially other geochemical testing methods are all examples of passive exploration methods. Identifying the geochemical and geophysical signatures that suggest the presence of the target REE-containing materials is usually the first step taken by the exploration members. Keep in mind, however, that the response of geophysical techniques (such as magnetometer, radiometric, gamma-ray spectrometry, and remote spectral surveys) depends upon the mineral associations in the rock that accompany the REE-containing mineral deposits. Detailed spectral data obtained with special spectroscopic methods from remote-sensing surveys can identify carbonate, ferrous iron, and REEs associated with thorium-rare earth element deposits (Armbrustmacher et al., 1995; Spellman, 2017).

The commonly identified and used passive explorations steps do not generate many wastes and have little if any environmental impact. The one exception is ground-based geophysical exploration, which may require clearing through forested or other heavy-growth areas to accommodate larger truck-mounted equipment, resulting in an alteration of the terrain. Also, earth-moving equipment may be used to remove soils and overburden or remove boulders that may be in the way of ground-survey activities. The spoil, soil and rock, removed during exploration activities may be stockpiled onsite and used for onsite purposes or reclamation activities, as needed. Ground surface disturbances are likely to continue as the exploration activity progresses. To further delineate and prove the resource using subsurface testing to sample, it is often necessary to construct additional access roads and drill sites

Exploratory Drilling and Trenching

An exploratory drilling program to further substantiate and quantify the attributes of the ore body or target deposit is accomplished using data obtained from prospecting activities, geophysical surveys, geochemical surveys, and geological modeling. These subsequent programs can incorporate various types of tests. The lateral and vertical extent of the REE deposit is determined by the drilling program used. Usually, rock cores are collected for further analysis to evaluate the continuity of mineralization, grade, mineralogical relationships, rock types, and local hydrogeologic data. Borehole geophysics and rock strength testing may also be performed, and trenching may be used to collect bulk samples consisting of several tons of material. This operation is accomplished across mineralized zones for metallurgical testing used to develop and evaluate processing methods that will be incorporated into the processing plant design. Large tunnels or large vertical shafts may be advanced to recover bulk samples using large earth-moving equipment.

Note that exploratory drilling can be ongoing for weeks, months, or even years. The actual footprint of a single drill site may be a few hundred square feet in size, and there can be numerous drill locations. These operations can potentially disturb and spoil several acres of land to accommodate the drill sites, staging areas, and the onsite support facilities for the drilling operation. The spacing of the drill site is dependent upon the continuity of the deposit. Exploratory borings may be drilled and spaced as close as 100 ft or less or at larger intervals, depending on the characteristics of the site (National Academy of Sciences, 1999). In probing, the REE deposits both waste solids and fluids are produced. As boreholes are advanced to greater depths rock flour and cuttings are removed. The volume and nature of cuttings produced at the ground surface depend upon the characteristics of the rock being drilled, the depth and diameter of the borehole, the presence of potable aquifers, and the drilling method. When using rotary drills it is common to use water and drilling mud (bentonite clays) as the drilling fluid to entrain and carry the cuttings to the surface while cooling the drill bit. Note that is common to use only water as a drilling fluid during exploration because clay drilling muds have the potential to complicate the petrologic and geochemical analyses; however, it is the formation characteristics and local hydrology that may make the use of drilling muds or other additives necessary. Compressed is used in other rotary drills to force the cuttings to keep the drill bit cool. Polymers, oils, or other synthetics are added to air or water flows to boost the operational properties of the drilling water or mud, if needed. It is important to note that prior to use, the drilling muds do not generally represent an environmental hazard and are either inert of break down quickly (Spellman, 2017; USEPA, 2012).

Note that when the borehole intersects with groundwater, the water can entrain materials from the borehole and transport those materials to the surface. This is an environmental concern, of course, because of the potential toxicity of these materials depends on the composition of the drilling fluids used and materials swept from the subsurface. These formation fluids can be highly mineralized, and the cuttings and drilling muds can entrain toxic levels of metals and other chemicals. Drill cuttings and drilling fluids are typically collected once the boreholes are completed and abandoned; however, regulations in some states allow return of cuttings to the borehole. Excess materials may be stored in drums, contained in mud pits, or managed and disposed as a wastewater or as a waste (Spellman, 2017).

Earlier in this book, the various locations where REE has been located and in some cases where it has been prospected within the United States were discussed. Well, redundancy is good (in my opinion) and in the following the various locations in the United States with their recent exploration activity and a summary of potential REE resource areas is pointed out—the keywords herein are "potential sites." Potential sites or REE resources have been under evaluation in recent years to prove the reserves and determine the mining potential at these sites. Sites receiving current attention (approximately around the 2011 timeframe) are located in Utah, Colorado, Wyoming, Alaska, California (current timeframe), Nevada, and Nebraska and provided here from USEPA (2012). Keep in mind that the below locations are only examples of sites where some level of exploration activities have occurred, but others may exist.

- During 2008, the Great Western Minerals Group, Ltd. completed an extensive drilling and sampling program at the deep sands deposit located in the Snake Valley and adjacent to the Deep Creek Mountains of west-central Utah. After a feasibility study to consider the low-grade monazite ore containing rare earth concentrations, the company announced that it did not plan to undertake additional exploration at the site (Great Western Minerals Group, 2011).
- Colorado Rare Earths Incorporated (CREI), a recently formed company, announced in March 2011 that it was soliciting funding to begin exploration near the Powderhorn mine site in Colorado (Colorado Rare Earths, Inc., 2011). CREI also has acquired properties at Iron Hill and Wet Mountains Colorado, where potential REE resources have been identified. CREI is currently organized as a mineral claims acquisition company but plans to expand to an exploration and development company in the future.
- In Alaska, exploration is underway at the Bokan Mountain site that is located at the southern-most end of the Alaskan panhandle on Prince of Wales Island. This is a site that was mined in the past for uranium ore, but also has known HREE and LREE resources.
- Molycorp Minerals is a past producer of REEs and is currently preparing to re-open its mine at Mountain Pass, California.
- In Nevada, about 16 mi. east of the Mountain Pass mine, Elissa Resources is exploring multiple locations called the Thor REE Project Area, where HREEs and LREEs have been discovered: at least one of these locations is geologically similar to Mountain Pass deposit (Elissa Resources, 2011). To date, Elissa Resources' exploration activities have included surface mapping, a district-scale detailed high-resolution airborne magnetic geophysical survey, extensive ground radiometric surveys, satellite imagery studies, satellite imagery studies, petrographic studies, and sampling program, including channel sampling and the sampling of historical surface workings that are remnants of a 1950s uranium–thorium prospected rush.
- Quantum Rare Earth Developments Corporation recently has begun exploratory drilling at the Elk Creek Project in southeastern Nebraska. The focus of these drilling efforts is to perform a resource assessment of niobium and associated REEs contained in carbonite deposits. These efforts are in the early stages, and no final determination about the resource has been made at this time (Quantum Rare Earth Developments Corporation, 2010).

- Bear Lodge Rare-Earth Project is located approximately 6 mi. north of Sundance, Wyoming and is comprised of 90 unpatented federal lode claims and one state lease, for a total of about 24,009 acres. It is owned by Rare Element Resources through its Wyoming-incorporated Paso Rico (USA) Company, Inc. Twenty-three of the mining claims were obtained from Phelps Dodge Corporation and are subject to 2% net smelter return production royalty. The site has been prospected at various times since 1949. Several areas of mineralization have been identified, including high-grade copper, molybdenum, gold, and REMs within an altered carbonatite-alkaline-intrusive complex. Note that the exploration efforts to date estimate that the rare earth resources at Bear Lodge could match or exceed the size and grade of deposits at Mountain Pass, California. The company web page reports that rare element resources are currently exploring for rare earths and gold, and results have been very good to date.
- Pea Ridge mine is an iron mine that has been in operation for about 40 years. The site byproduct from the iron ore body; however, mining of an adjacent ore body containing primarily REEs had also been planned (Wings Enterprises, 2011).

Mine Development

A financial analysis and corresponding feasibility study are important parts of this stage, as defined here, and are conducted considering the market value of the metal commodity, cost of production characteristic of the deposit, and anticipated closure costs. The results are then used to determine if the deposit is not likely to ever be an economically viable deposit or if it could become economically viable in the future, or if the development of the deposit is economically feasible and should proceed. A feasibility study generally takes about three to four months for a medium-size mine and, for a larger project six to nine months (De la Vergne, 2003). In mine development, the environmental effects of the proposed mining activity must be considered along with long-term costs. The potential impacts to the immediate property, adjacent lands, and surrounding community must also be identified and considered by performing additional baseline environmental studies. The results of these preliminary studies can be used to convince outside investors to sign on to the project.

Upon completion of the feasibility study that is positive, putting together the mine plan is typically next on the to-do list. The layout and design of the mine are performed to locate stockpile areas and waste areas where there can be best managed to prevent or minimize environmental damage. It is prudent to and cost-effective (best practice) to conduct the engineering design and EA together so that collected and measured data can mutually benefit both purposes and allow for the collaboration of engineers and environmental scientists. When design engineers do not consult with environmental professionals throughout the planning and design phase the end result can be quite costly. It is more cost-effective (and easier) to plan for environmental requirements (including safety and health of all involved) early on and throughout the planning phase than it is to find out later that some environmental regulatory requirement or safety and health issue was overlooked—overlooking environmental requirements can be very expensive. For example, many mining sites have storage areas for chemicals used in the mining process. Although these chemical storage areas or structures are usually temporary and only used when mining is actively being conducted, there are toxic hazards within the chemical storage area/structure that workers handling the chemicals can be exposed to and contaminated by. Safety regulators like MSHA (Mine Safety and Health Administration) require that eye wash/deluge shower equipment to be installed inside or within a few feet of the structure outdoors—for easy and prompt access in case of employee contamination. Many operations are aware of this requirement and often opt for portable eyewashes that contain a small amount of water to flush the eyes. However, because many of the chemicals used in mining are hazardous when spilled on the body, deluge showers that can provide a constant flow of water when needed are required. If these eyewash/deluge showers are part of the initial planning process this system can be part of the initial staging process, which in the long run is less expensive to install initially than it is to install as an afterthought—after an MSHA fine, or worse, for example. In all of the years of my professional environmental engineering work I have lived by my safety and health mantra, which is

> In the world of safety and health of workers mitigating a hazardous situation before it occurs is 1,000 times better than litigation after the hazardous event occurs.

When we speak about mining accomplished without leaving a huge, disastrous footprint it is wise to remember that the footprint can be hazardous to the miners and others, as well. Simply stated mitigation measures can be addressed more efficiently though management plans and compensation measures that are intreated into the feasibility models to evaluate cost—both monetary and human costs.

As pointed out earlier, REEs commonly occur in ores that also contain uranium and thorium. Ores containing the REE-bearing mineral monazite are particularly high in concentrations of thorium (Long et al., 2010). Uranium also occurs in or with REE-ore deposits. The amounts of radioactive elements in the ore are not dependent upon the mineral type specifically, but more on the petrogenesis of the deposit containing the ore. In general, nearly all rocks, soils, and water contain small amounts of radioactive material, such as uranium, thorium, radium, radioisotopes of potassium, lead polonium, and their decay products. When naturally occurring radioactive materials (NORM), in their undisturbed natural state, become inadvertently or purposefully concentrated, either in waste by-products or in a product, they become technologically enhanced naturally occurring radioactive materials (TENORM) (USEPA, 1999).

Whenever waste areas contain high concentrations of uranium and thorium complex management is involved and required. Management begins at the mine design stage, where the mining methods and milling plant processes are planned to optimize a waste reduction-minimization strategy. Again, safety and environmental issues must be front and center in all mine planning efforts especially in regard to the treatment of radioactive tailings. The EPA reports (USEPA, 1999) that the radiation levels from waste rock and sludges associated with the production of REOs range from 5.7 to 3,224 pCi/g (Average Picocuries Per Gram). Additional treatment circuits are generally required to precipitate radium from tailings, and additional controls are needed to manage radon. Production and management of radioactive mining and milling wastes are tightly controlled by national (including the Safe Drinking Water Act, Clean Water Act, Comprehensive Environmental Response, Compensation and Liability Act [CERCLA], and Clean Air Act) and state regulations, and if REE is produced with commercial uranium or thorium, then international protocols and agreements also may apply. The challenge is generally balancing the human health and environmental risks with production costs. The issues associated with these materials can affect the feasibility and viability of a proposed mining project, especially whenever planning to reopen an old mine site. Tighter regulation on the use of radioactive minerals is the primary factor that pressured many sources of monazite out of the REEs market during the 1980s (Long et al., 2010).

Processing ores containing radionuclides subsequently increases the concentration of naturally occurring radioactive, materials and the levels of radioactivity can become higher than the background levels near background waste areas. However, it is important to point out that radioactive wastes from mining and milling operations are different from waste from enriched radioactive materials generated by nuclear fuel facilities or disposed of by nuclear power plants. At mining and milling sites, it is the volume of waste containing radioactive elements that are the main concern. With regard to protecting the environment, if mines are located on or adjacent to federally managed lands or will have discharges to surface water then an Environmental Information Document (EID) (aka Environmental Study or Report) is required under NEPA. Preparation of this document is usually performed by the proponent during early mine stage activities as the viability of the mining operations is being assessed. The EID is important because it describes the project, characterizes the environment that potentially may be impacted by the mining activities, assesses the potential severity of any impact, identifies mitigation measures to avoid or lessen the impacts (remember mitigation in the beginning may avoid litigation in the end), and discusses alternatives for methods and operations, that may include an evaluation of the critical need of the commodity (e.g., availability of the commodity from other markets, substitute materials, or recycling). Information contained in the proponent's EID is used by the managing agency for the project to prepare and EA report describing the potential for environmental impact and the preferred alternative. If the mine project is considered a major federal action that has the potential to significantly affect the quality of the human environment, then the EID and/or EA is used by the lead government agency to develop an Environmental Impact Statement (EIS); still, additional environmental baseline and engineering studies may be needed. The EIS is then reviewed by EPA and other agencies, if applicable, as part of the NEPA process to determine if unavoidable adverse impacts will occur, and if irreversible and irretrievable commitments of resources are likely, considering connected actions, cumulative actions, and similar actions. A period of public review and the issuance of a Record of Decision (ROD) follows. If an EIS is implemented, monitoring is required for any mitigation steps required and implemented.

Note that in the environmental impact investigation(s) it is determined what the cumulative impacts of the mining and processing activity might entail. In this regard keep in mind that common baseline studies performed to support the EID might include an evaluation of aquatic resources (including wetlands), terrestrial habitat, groundwater quality and supply, air quality, and human population and demographics. Each of these studies focuses on characterizing the environment that potentially may be affected by the operations and activities at the site. The information from baseline studies is then used in the EID to assess the potential impact that the operations and activities may have on the site and surrounding environment, relative to the potential impacts identified and anticipated for the mining method used and the associated operations and processes. The impact may include degradation of habitat or habitat alteration or loss; on the other hand, in some special cases, habitat may actually be enhanced or created. Water use for mining operations may also cause flow alterations that have adverse impacts to aquatic habitat, or in-stream structures used for water control may obstruct fish movement, migration, and spawning patterns. An aquatic resource study generally evaluates the distribution, abundance, and condition of fish species, amphibians, and benthic macroinvertebrates, along with the distribution and extent of the habitat and riparian zones.

DID YOU KNOW?

Freshwater macroinvertebrates are ubiquitous: even polluted waters contain some representatives of this diverse and ecologically important group of organisms. Benthic macroinvertebrates are aquatic organisms without backbones that spend at least a part of their life cycle on the stream bottom. Examples include aquatic insects—such as stoneflies, mayflies, caddisflies, midges, and beetles—as well as crayfish, worms, clams, and snails. Most hatch from eggs and mature from larvae to adults. The majority of the insects spends their larval phase on the liver bottom and, after a few weeks to several years, emerges as winged adults. The aquatic beetles, true bugs, and other groups remain in the water as adults. Macroinvertebrates typically collected from the stream substrate are either aquatic larvae or adults (Spellman, 2019).

To assess existing stress on the aquatic environment, water quality criteria are evaluated against the actual condition of the water body, and toxicity studies and metric analysis of macroinvertebrates may also be performed. Aquatic impacts from mining and mine processor are often considered to be the most significant.

Again, because aquatic impacts of mining are so significant a metric analysis of macroinvertebrates is a relatively easy way in which to monitor mining impact on local streams.

So, the obvious question is what is a metric analysis of macroinvertebrates?

Well, one of the best ways to answer this question is to provide an example; for our purposes the biotic index is included and described herein.

Simply, the biotic index is a systematic survey of macroinvertebrates organisms. Because the diversity of species in a stream is often a good indicator of the presence of pollution such as mining waste, the biotic index can be used to correlate with stream quality. Observation of types of species present or missing is used as an indicator of stream pollution. The biotic index, used in the determination of the types, species, and numbers of biological organisms present in a stream, is commonly used as an auxiliary to biological oxygen demand (BOD) determination in determining stream pollution. The biotic index is based on two principles:

1. A large dumping of organic waste into a stream tends to restrict the variety of organisms at a certain point in the stream.
2. As the degree of pollution in a stream increases, key organisms tend to disappear in a predictable order. The disappearance of particular organisms tends to indicate the water quality of the stream.

Several different forms of the biotic index. In Great Britain, for example, the Trent Biotic Index (TBI), the Chandler score, the Biological Monitoring Working Party (BMWP) score, and the Lincoln Quality Index (LQI) are widely used. Most of the forms use a biotic index that ranges from 0 to 10. The most polluted stream, which therefore contains the smallest variety of organisms, is at the lowest end of the scale (0); the clean streams are at the highest end (10). A stream with a biotic index of greater than 5 will support game fish; on the other hand, a stream with a biotic index of less than 4 will not support game fish.

As mentioned, because they are easy to sample, macroinvertebrates have predominated in biological monitoring. Macroinvertebrates are a diverse group. They demonstrate tolerances that vary between species. Thus, discrete differences tend to show up, containing both tolerant and sensitive indicators. Macroinvertebrates can be easily identified using identification keys that are portable and easily used in field settings. Present knowledge of macroinvertebrate tolerances and response to stream pollution is well documented. In the United States, for example, the Environmental Protection Agency (EPA) has required states in incorporate narrative biological criteria into its water quality standards by 1993. The National Park Service (NPS) has collected macroinvertebrate samples from American streams since 1984. Through their sampling effort, NPS has been able to derive quantitative biological standards (Spellman, 2019).

The biotic index provides a valuable measure of pollution. This is especially the case for species that are very sensitive to lack of oxygen. An example of an organism that is commonly used in biological monitoring is the stonefly. Stonefly larvae live underwater and survive best in well-aerated, unpolluted waters with clean gravel bottoms. When stream water quality deteriorates due to organic pollution, stonefly larvae cannot survive. The degradation of stonefly larvae has an exponential effect upon other insects and fish that feed off the larvae; when the stonefly larvae disappears, so in turn do many insects and fish (Spellman, 2019).

Table 9.1 shows a modified version of the BMWP biotic index. Considering that the BMWP biotic index indicates ideal stream conditions, it considers the sensitivities of different macroinvertebrate species are represented by diverse populations and are excellent indicators of pollution. These aquatic macroinvertebrates are organisms that are large enough to be seen by the unaided eye. Moreover, most aquatic macroinvertebrates species live for at least a year; and they are sensitive to stream water quality both on a short-term basis and over the long term. For example, mayflies, stoneflies, and caddisflies are aquatic macroinvertebrates that are considered

TABLE 9.1
BMWP Score System (Modified for Illustrative Purposes)

Families	Common-Name Examples	Score
Heptageniidae	Mayflies	10
Leuctridae	Stoneflies	9–10
Aeshnidae	Dragonflies	8
Polycentropidae	Caddisflies	7
Hydrometridae	Water Strider	6–7
Gyrinidae	Whirligig beetle	5
Chironomidae	Mosquitoes	2
Oligochaeta	Worms	1

clean-water organisms; they are generally the first to disappear from a stream if water quality declines and are, therefore, given a high score. On the other hand, tubificid worms (which are tolerant to pollution) are given a low score.

In Table 9.1, a score from 1 to 10 is given for each family present. A site score is calculated by adding the individual family scores. The site score or total score is then divided by the number of families recorded to derive the Average Score Per Taxon (ASPT). High ASPT scores result due to such taxa as stoneflies, mayflies, and caddisflies being present in the stream. A low ASPT score is obtained from streams that are heavily polluted and dominated by tubificid worms and other pollution-tolerant organism. From Table 9.1, it can be seen that those organisms having high scores, especially mayflies and stoneflies, are the most sensitive, and others, such as dragonflies and caddisflies, are very sensitive to any pollution (deoxygenation) of their aquatic environment.

As noted earlier, the benthic macroinvertebrate biotic index employs the use of certain benthic macroinvertebrates to determine (to gauge) the water quality (relative health) of a water body (stream or river). Benthic macroinvertebrates are classified into three groups based on their sensitivity to pollution. The number of taxa in each of these groups is tallied and assigned a score. The scores are then summed to yield a score that can be used as an estimate of the quality of the water body life.

Metrics within the Benthic Macroinvertebrates are best explained in the following:

Table 9.2 provides a sample index of macroinvertebrates and their sensitivity to pollution. The three groups based on their sensitivity to pollution are as follows:

Group One—Indicators of Poor Water Quality
Group Two—Indicators of Moderate Water Quality
Group Three—Indicators of Good Water Quality

In summary, it can be said that unpolluted streams normally support a wide variety of macroinvertebrates and other aquatic organisms with relatively few of any one kind. Any significant change in the normal population usually indicates pollution.

TABLE 9.2
Sample Index of Macroinvertebrates

Group One (Sensitive)	Group Two (Somewhat Sensitive)	Group Three (Tolerant)
Stonefly larva	Alderfly larva	Aquatic worm
Caddisfly larva	Damselfly larva	Midgefly larva
Water penny larva	Cranefly larva	Blackfly larva
Riffle beetle adult	Beetle adult	Leech
Mayfly larva	Dragonfly larva	Snails
Gilled snail	Sowbugs	

Okay, let's shift back to milling and processing involved in mining. These processes are sometimes shared by multiple nines, and ore must be conveyed to the mill or processor site. The impacts on health and safety from the method used for ore conveyance, often by heavy truck or railway, through in-route communities and sensitive areas must be considered in relation to existing traffic patterns, spillage of loads along routes, and the degradation of air quality. This generally includes the release of windblown air particulates and vehicle exhausts. However, in addition to mobile sources of air emissions that could emit any of the hazardous air pollutants (HAPs) that are regulated under the Clean Air Act (CAA). HAPs are not covered by ambient air quality standards, but which (as defined in the Clean Air Act) may reasonably be expected to cause or contribute to irreversible illness or death. Such pollutants include asbestos, beryllium, mercury, benzene, coke oven emissions, radionuclides, vinyl chloride, and mining emissions. A list of HAPs is provided in CAA 112.b.

In order to determine the severity of potential impacts to nearby human receptors, a population study is usually performed. In general, a population study is a study of a groups of individuals taken from the general population who share a common characteristics, such as age, sex, or health condition. This group may be studied for different reasons, in the case of populations related to, nearby mining operations a land-use mapping technique, also called analysis, is used to aid with population studies and to help identify farming areas, parks, schools, hospitals, and other potentially sensitive areas (both human and ecological habitat) adjacent to the proposed mine site. Evaluation of local demographics is generally performed to insure fair treatment of racial, ethnic, and socioeconomic groups.

EPA stipulates that three months may be needed for the review of an EID, but a longer period may be necessary or required. NEPA also describes the preparation of the EIS and the intra-agency review process as potentially requiring 12 months or longer. Consequently, there could be a period of several years between initial exploration steps and startup of a REE mine that is located on federally managed lands. If needed, land, mineral leases, and all regulatory permits would be acquired pending acceptance of the EIS. If environmental audits are required they will also be performed.

In 2011, the United State Geological Survey evaluated the time it took to develop mines that opened in the United States since 2000. It was found that even if expedited permitting occurs, the period to obtain a permit could take seven years. It was also found that periods of one month to 17 years may be required before commercial production. Ramp-up times of new mines might take from 2 to 12 months.

The typical timeframe for a mine project from prospecting (i.e., initial exploration) through the construction of the mine was generalized by USEPA Region 10 and is presented in Table 9.3, below (USEPA, 2011).

The reality is that there are numerous factors that affect the time required for mine development. Some examples might include the location of the mine (e.g., federal lands, state lands, or private lands and greenfields versus brownfields); geologic complexity of the site; metallurgical complexity of ore; level of stakeholder involvement; availability of required infrastructure; and experience of regulatory authorities with mine type and area deposit is located (USEPA, 2012).

In addition to conducting environmental due diligence to ensure that the mining operations will not produce irreputable harm to the environment, humans and wildlife mining companies must also provide for financial assurance that the mine site will be restored to a condition that does not represent a risk to the environment or human health. The federal Superfund law, also known as CERCLA, contains a provision that requires EPA to establish financial assurance requirements for facilities that produce, treat, store, or dispose of hazardous substances. Under CERCLA, a wide range of elements, compounds, and waste streams are specifically designated as "hazardous substances" in Title 40 of the Code of Federal Regulations, Section 302.4. Many states have established financial assurance requirements that are in place; however, the CERCLA 108(b) financial assurance regulation for hard rock mining currently implements section 108(b) of CERCLA (43 U.S.C. 9608(b)(1), which directs the President to promulgate requirements that

> ... classes of facilities establish and maintain evidence of financial responsibility consistent with the degree and duration of risk associated with the production ... treatment, storage or disposal of hazardous substances.

TABLE 9.3
Typical Timeframe for a Mine Project

Mine Project	Timeframe in Year
Initial exploration	1 to 2
Advanced exploration	2 to 4
Environmental studies	2 to 7.5
Prefeasibility studies	3.5 to 5.5
Feasibility studies	5.5 to 7.5
NEPA and Permitting	6.5 to 8.5
Financing	8.5 to 10.5
Construction	10.5 to 12.5

The EPA has worked on a methodology that will be used to determine the level of financial responsibility requirements that will be imposed on the mine or mine facility owner under Section 108(b) of CERCLA.

EPA's responsibilities in this area are defined by the federal Superfund law (CERCLA).

CERCLA contains provisions that give EPA authority to require the classes of facilities to maintain financial responsibility consistent with the degree and duration of risk associated with the production, transportation, storage, or disposal of hazardous substances.

Because rare earths are often a constituent in ores processed to recover other metal or mineral commodities this may not only create the opportunity to expand current mining activities in some locations but also provide a means to reduce costs due to investment and financial backing already available because of the current mining operation. Indeed, the demand for rare earths is increasing daily and this may create opportunity for operating mines to consider mining rare earths for additional profit—the old two valuable ores or more with one action, so to speak. However, keep in mind that expansion of mining operations and changes to milling and processing operations may require environmental review and additional permitting for these active mining operations (USEPA, 2012).

In addition to REE production from current mining sites, it could become profitable for operations to resume at former mine locations. These may be mines with active permits that are currently not in operation for one reason or another but usually because the current market value of the principal commodity mined, or closed mines where ore remains but no mining is occurring, the equipment has been removed or mothballed on site, and possibly some level of reclamation has been started or completed. One source of REE may be present in old stockpiles, waste tailings, which may represent a potential windfall that can help the mine transition back into production as the mine is reopened. Most of the prospective REE mines in the United States include those that produced REE ores in the past at a time when they were not in as great as demand as they are currently—the transition from just REE to white gold, you could say. A good example of this type of mine is Molycorp Mineral's mine in Mountain Pass, California and also a mine that in the past produced another commodity that contained REEs like the Pea Ridge iron ore mine. While it is true that some exploration may be required it is also true that enough existing data may be available to begin planning the re-opening of the mine. Also, keep in mind that most of the activities required for the mine development stage will be needed to reopen the mine. In some cases, the environmental impacts from past mining activities and practices will need to be considered in the development of the new mining operation.

Ore Extraction and Processing

Upon receiving regulatory approvals have been acquired, construction can begin on waste management and processing areas, followed by the commencement of mining activities. The mining methods that might be used to recover REE ores are similar to hard rock or placer mining operations used for extracting other metal ores.

Hard rock mines are typically either aboveground or underground operations. Solution mining techniques could also be used to extract REM oxides from the subsurface however, the use of solution mining is unlikely in the United States due to the potential for environment impacts and the absence of suitable deposits. Placer deposits are typically near-surface deposits that extend over great areas rather than to great depths.

Methods used to extract ore are dependent on the grade of the ore; size of the deposit; ore body position (i.e., the size, shape, dip, continuity, and depth); geology of the deposit; topography; tonnage, ore reserves; and geographic location and is based on maximin gore recovery within economic constraints. The ore extraction methods currently used are discussed in detail in Chapter 10 of this book for now let's just point out that a low-grade cutoff point is established on a site-specific basis and depends on recovery costs at the site, the market prices of the ore, and feed requirements at the mill. It is common for open-pit and underground mining to occur at the same site. Aboveground mind pits are generally conical in shape, with the diameter decreasing with increasing depth; however, a conical pit is not always developed, and the advance of some aboveground mines is controlled by topographic and/or geologic features. As the mining proceeds, there may no longer be adequate room for equipment to work safely in the pit, and it may not be feasible or practical to remove additional waste rock to widen the pit. After that point, an underground mine may be developed inside the lower reaches of the mine pit to reach deeper parts of the deposit. The economics of widening the pit may also make mining of deeper sections of the ore body using underground methods more advantageous.

Mine Closure

Simply stated the closure of a mine refers to the cessation of mining at that site. Not so simple are the steps needed to properly close the mine. For example, closing the mine involves completing a reclamation plan and ensuring the safety of areas affected by the operation including sealing the entrance to an abandoned mine. EPA requires that planning for closure is ongoing for mines located on federal lands and not left to be addressed at the end of operations. The Surface Mining and Control Act of 1977 states that reclamation must

> restore the land affected to a condition capable of supporting the uses which it was capable of supporting prior to any mining, or higher or better uses.

In some cases, reclamation may not be possible to a level that would support past uses, and long-term monitoring and management may be required (USEPA, 2012).

Note that mines close for a variety of reasons, including economic factors due to the decline in the market value of the metal; geological factors such as a decrease in grade or size of the ore body; technical issues associated with geotechnical issues, such as mine stability or in-rush of groundwater at rates too high to manage; equipment failure; environmental impacts, and community pressures (Lawrence, 2003).

While some mines do close for any of the reasons listed above but there are mines that are idled and not closed with plans to reopen when the market will support the venture. The problem with idle mines is that they can be a repository of on-going health-related hazards and threats to the environment, especially if not properly managed. While the mine is idle, the waste piles, the tailings ponds that are associated with the processing plant, and other mine areas must be stabilized and managed, potentially over many years.

Experience has shown that mines can be adequately managed to avoid long-term problems; however, the pollution and contamination issues that generally require environmental management over the life of the mine can remain after closure, such as the following (USEPA, 2012):

- sedimentation of surface waters
- effluent and drainage (e.g., due to heavy precipitation during the rainy season and snow melt) from the min storage, piles, or tailings management areas that can impact downstream drinking water sources and adequate habitat
- acid mine drainage (AMD) and enhanced acid rock drainage (ARD) that also effect aquatic habitats
- continued pumping of mine water and contaminant plume migration control pumping that causes drawdown of nearby aquifers
- contaminate dusts
- subsidence or collapse of tunnels and subsurface structures, which also represent a safety hazard
- hazardous materials remaining on-site (e.g., fuels, lubricants, other chemicals).

These management activities must be planned for long after operations cease and the mine is closed. The associated with these impacts are discussed in Chapter 13.

Note that it is the state or a land management agency that has the final authority for concerting regulatory authorities and overseeing mine closures on public lands. For instance, EPA's responsibilities may include but not be limited to NPDES wastewater discharge permits for post-closure mining operations, Clean Water Act (CWA) and Resource Conservation and Recovery Act (RCRA) related permits required for wastewater or contaminant treatment options (if necessary), and CERCLA financial responsibility documentation.

The bottom line: at the present time we do not know what we do not know about potential future applications of REEs. One thing we do know for certain is that REEs are becoming important and critical on a global scale—meaning that more mining for REEs is inevitable.

REFERENCES

Armbrustmacher, T.J., Modreski, P.J., et al. (1995). Mineral deposits models: the rare earth element vein deposits. United State Geological Survey. Accessed 12/28/2021 @ http://www.pubs.usgs.gov/of/1995/ofr-95-0831/CHAP7.pdf.

Colorado Rare Earths, Inc. (2011). Company webpage. Accessed 12/28/2021 @ http://www.coloradorareearth.com/news.htm.

De la Vergne, J. (2003). *Hard Rock Miner's Handbook*, 3rd ed. McIntosh Engineering, Temper, AZ. Accessed 12/28/2021 @ http://www.world-of-stocks.com/hp_contgnet/pdf/Mcintosh%20Engineerign_rulesofthump.pdf.

Elissa Resources (2011). Company internet page. Accessed 12/28/2021 @ http://www.elissaresources.com/projects/rare_earth/thor_ree_project.

Great Western Minerals Group (2011). Deep Sands, Utah, rate earth element project. Accessed 12/28/2021 @ http://www.gwing.c/html/projects/deep-sand/index.cfm.

Lawrence, D. (2003). Optimization of the mine closure process. School of Mining Engineering, University of New South Wales. *Journal of Cleaner Production* 14: 285–298.

Long, K.R., Bradley, S., et al. (2010). The principal rare earth elements deposits of the United States—a summary of domestic deposits and a global perspective. USGS Scientific Investigation Report 2010–5220, Reston, Virginia. Accessed 12/28/2021 @ http://usgs.gov/sir/2010/5220/.

National Academy of Sciences (1999). *Hardrock Mining on Federal Lands*. Committee on Hardrock Mining on Federal Lands, National Research Council. Washington, DC: National Academy Press.

Quantum Rare Earth Development (2010). Rare earth development. Accessed 12/12/21 @ https://www.prweb.com/releases/2010.

Spellman, F.R. (2017). *The Science of Environmental Pollution*, 3rd ed. Boca Raton, FL: CRC Press.

Spellman, F.R. (2019). *Water and Wastewater Operators Handbook*, 4th ed. Boca Raton, FL: CRC Press.

USEPA (1999). Technical report on technologically enhanced naturally occurring radioactive materials in the southwestern copper belt of Arizona. EPA Document 402-402-R-99-002 Office of Radiation and Indoor Air, Radiation protection Division, Washington, DC. Accessed 12/28/2021 @ http://www.epa.gov/radiation/docs/tenorm/402-r-99-002.pdf.

USEPA (2008). Ore mining & dressing effluent guidelines. Accessed 12/24/2021 @ https://www.epa.gov/eg/ore-mining-and-dressing-effluent-guidelines.

USEPA (2011). Mining information session part 1: mining fundamentals (representation). U.S. EPA Region 10, State of Alaska, June 2. Accessed 12/28/2012 @ http://www.epa.gov/region 10pdf/bristolbay/mining_session_presentation_part2.pdf.

USEPA (2012). Rare earth elements: a review of production, processing, recycling, and associated environmental issues. EPA 600/R-12/572. Accessed 12/23/2021 @ www.epa.gov/ord.

Wings Enterprises. (2011). Company webpage: press release January 2011. Accessed 12/28/2021 @http://www.wingsironore.com.

10 Excess and Unused Materials

ENVIRONMENTAL RISKS OF MINING REE

"Excess risks and unused materials?"—what does that mean? Well, it simply means that when we disturb our natural surroundings by mining (or by any other way or means, for that matter) there are consequences. The consequences are many and varied, local, regional, global, and direct or indirect or all of those listed herein. Mining can result in air pollution via carbon emission erosion, loss of biodiversity, sinkholes, contamination of groundwater, and surface water by mining wastewater that enters lakes, rivers, or streams. As a case in point considers Case Study 10.1.

Case Study 10.1—A Personal Observation: When the Animas River Became a Yellow Boy

In August 2015, near Silverton, Colorado, after the EPA discharged an estimated 3 million gallons of contaminated water from the Gold King mine into the Animas River (New Mexico) and created a toxic sludge, officials with the Navajo Nation, San Juan County, and New Mexico officials were shocked. Having personally observed this pollution event while in the Cement Creek area (where the spill originated; see Figures 10.1(a)–(e) during that timeframe, I suffered from what I call the Double D Syndrome: Disgusting and Disturbing.

DOI: 10.1201/9781003350811-12

FIGURE 10.1 (a) Rough illustration of spatial relationship of Gold King Mine to Animas, San Juan, Colorado Rivers and Lake Powell. (b) Silverton, Colorado area. USGS public domain photo accessed from https://usgs.gov/media/images/panorama/silverton-colorado. (c) Entrance to Gold King mine. USEPA public domain photo accessed from https://www.epa.gov.goldkingmine. (d) Shows Gold King mine wastewater holding ponds. USEPA public domain photo accessed from https://www.epa.gov.goldkingmine. (e) Shows worker adding lime to Yellow Boy (Animus River—King Gold mine spill). USEPA public domain photo accessed from https://www.epa.gov.goldkingmine.

What many do not understand about this particular pollution event is that the toxic cocktail that makes up the contaminated sediments will eventually settle in the river bottom or bed. When this occurs, the sediments do not magically disappear. Hercules was said to have stated that dilution is the solution to pollution. To a degree this is true, of course, but generally is only true to suspended pollution contaminants, sediments, again, settle. Thus, it is difficult to dilute that which does not go away; those contaminants that persist. The problem is that future runoff from storms will lift the toxic sediments from the river bed back into the water, which means the contamination is long-term; almost chronic for decades and even longer.

Another problem compounding this latest pollution event is that the Animas River has not been exempt from previous acid mine drainage events. On the contrary, in the past acid mine drainage into the Animas River has been on-going is well documented.

The bottom line: "The spill was loaded with heavy metals, including arsenic, lead, copper, aluminum, and cadmium and turned the Animas River into the Mustard-colored River—what the old-time miners call a Yellow Boy."

Case Study 10.2—Acid Rock Drainage/Acid Mine Drainage

The chemical interactions between rock, soil, and water cause minerals to leach out of rock and into soil, surface water, and groundwater. Water sources and soil in areas of heavy mineralization where water travels over ore deposits can be heavily affected by mineralization, either naturally or through the eventual mining of the ores. In fact, mining sites are sometimes discovered through the surficial signs of natural acid rock drainage (ARD). Human activities, including building and highway construction, mining, quarrying, and logging can increase the amount of mineral released into soil and water supplies.

Biota in areas of high mineralization is affected by leachates with high metal toxicity and sulfide-bearing materials. Metal tolerant plants grow in such areas, and the plant life on the fringes of these areas eventually overgrow acidic "kill zones." Aquatic life is less adaptive, however, and creeks and lakes with high acid concentrations are often devoid of life, whether through natural ARD or acid mining drainage.

While ARD and acid mining drainage produce similar effects, establishing a baseline for natural acid drainage allows geologists to plan mine development and predict and treat the eventual mine drainage (Spellman, 2017).

Note: Because some mining methods such as REE mining may have significant environmental and health effects that mining entities in most countries, including the United States, must abide by and follow strict environmental and rehabilitation codes to ensure that the mined area returns to its original state.

NOT SO GREEN

Today, there is a serious movement toward going "green." Not talking about going to and the accumulation of and the taking in greenbacks (the almighty dollar, pound, etc. but instead about shifting from polluting energy sources to those so-called green energy sources (e.g., wind power, solar power, power cell power, battery power, hydraulic power, ocean power, and even nuclear power) to replace hydrocarbon pollution sources.

Hmmmm! Sounds like a good idea, doesn't it? A panacea for the future. Well, it is to a point. Anytime we can continue the so-called "good life" without polluting our environment that is a good thing. The truth be known there is a complicated legacy of REE mining. Let's take a look, a closer look, at this complicated legacy of REE mining.

Okay, to start with we know we need REE for new technological developments that will allow us to swift from fossil fuels to renewable energy all provided by electricity and all its benefits like for powering electric vehicles (EVs) which do not pollute our environment and also there are the wind turbines and solar cells that provide pollution-free energy—we know all this. However, keep in mind that REE is used extensively in these pollution-free systems but has to be mined. If the mining is unregulated or not monitored, it has the potential to release a harmful substance in the air, soil, and water. Let's look at some of these pollution sources—sources that make the mining of REE and/or anything else without consequences.

Open Pit Mining

When the material is excavated from an open pit, it is called open pit mining. Open pit mining is one of the most common forms of mining for strategic, tactical, and/or planned minerals. Open pit mining is particularly damaging to the environment because REE are often only available in small concentrations, which is increases the amount material (ore) needed to be mined. The environmental hazards are virtually present in every stop of the open-pit mining process. Keep in mind that hard rock mining exposes rock that has lain unexposed for geological eras. When crushed, REE may expose radioactive elements, asbestos-like minerals, and metallic dust. While in the separation process, residual rock slurries, which are mixtures of pulverized rock and liquid, are produced as tailings, radioactive elements these liquids can leak into bedrock if not properly contained.

Okay, open pit mining of REE has its problems, how about underground mining?

Underground Mining

From a safety prospective, underground mining has the potential for deadly collapses and land subsidence (Betournay, 2011). It involves large-scale movements of vegetation and waste rock, comparable to open pit mining. In addition, like other forms of mining, underground mining can release toxic materials into the soil, air, and water. *Mining waste* (wastes from mining, milling, smelting, and refining ores and minerals) is a considerable contributor to soil contamination. Environmental problems are primarily associated with the disposal of mining waste: the overburden plus *tailings* (residue of ore processing). If piled into heaps and left at the site, rainwater seeps through mine wastes, which are rich in heavy metals and chemicals, and harmful byproducts like sulfuric acid may be produced. This acid runoff then drains into rivers, streams, soils, and then groundwater. In addition, because mining generally involves tapping into subterranean sources of water, and/or water is used in some fashion (hydraulic pumping and suction dredges) during mining the used water has to go somewhere and it does. The problem is that as the water flows wherever it can during digging operations it picks up and transports contaminants such as chemicals used in the mining process and heavy metals in the mine waste. This contaminated wastestream can pollute the area surrounding the mine and also pass the contaminated wastestream into other locations and it may enter groundwater or a local stream, river, or lake (Spellman, 2017). In addition to contaminating neighboring bodies of water the mining operation along with the flow of wastewater removes topsoil making it difficult for plant life to recover. In the case of mining in forest areas, valuable forest biomes are destroyed and the landscape is exposed to erosion.

Other Mining Techniques

In addition to open pit and underground mining other mining techniques are used to retrieve important minerals from the ground. These techniques include in situ leach (ISL), heap leaching, and brine mining. *ISL mining* has both environmental and safety advantages of conventional mining in that the ore body is dissolved and then pumped out, leaving minimal surface disturbance and no waste rock or tailings (World Nuclear Association, 2012). However, the strong acids used to dissolve the ore body commonly dissolve metals in the host rock as well. The fluids left behind after the leaching process commonly contain elevated concentrations of metals and radioactive isotopes, posing a significant risk to nearby ground and surface water sources (IAEA, 2005).

The *heap leaching* mining process involves employing a heap leaching circuit whereby a series of chemical reactions absorb specific minerals and separate them from other earth materials. The environmental issues involved with heap leaching are centered on the failure to keep process solutions within the heap leaching circuit. When heap leaching fluids are released into the environment they can affect the health of both the surrounding ecosystem and human population (Reichardt, 2008). In the *brine mining* process, extraction and evaporation of brine solutions are used to remove harmful elements and compounds (Gruber et al., 2011), potentially released them into the environment.

The bottom line: Mining for REE must be conducted in a safe, sound, and environmentally favorable manner. Otherwise, excess and unused materials that out-and-out contaminants can be released into the soil, the air, and our water supplies.

MINING WASTE CHARACTERISTICS

When it comes to defining waste there are all kinds of opinions and definitions about what waste actually is. As described in my *Science of Waste* book, "waste" is defined as something no longer wanted, something destroyed or broken or damaged beyond repair and therefore disposed of or simply thrown away because it is no longer functional, no longer needed, and/or no longer wanted. Well, when it comes to mining waste we have to reload, so to speak, because mining waste could be referred to as managed materials of concern and, in general, some materials are not considered wastes until a particular time in their life cycle. However, that is not to imply an absence of environmental risk from stockpiled or stored materials. Mined materials that are generated may only occasionally or periodically be managed as wastes. Frequently, these materials are used for various onsite or offsite purposes instead of being stored as wastes, although the volume of waste material can often exceed the demand for reused alternatives.

EPA evaluates the risks to human health and the environment from the reuse of various types of industrial sector wastes, including mining; however, many of these studies are incomplete, and we are concerned here with reuses or reuse issues as they relate to mining wastes from hard rock mining operations (e.g., REE mines).

The areas or practices of concern related to mining waste are mining-influenced water (MIW) and its treatment, soil storage piles, overburden, ore and subeconomic ore storage, and waste rock.

MINING-INFLUENCED WATER (MIW)

MIW is defined as any water whose chemical composition has been affected by mining or mineral processing and includes ARD, neutral and alkaline waters, mineral processing waters, and residual waters. MIW can contain metals (including REEs), metalloids, and other constituents in concentrations above regulatory standards. MIW affects over 10,000 mi. of receiving wastes in the United States, primarily from acid drainage. In this book, a brief overview of select MIW treatment technologies used or piloted as part of remediation efforts at mine sites. The 16 technologies listed and described herein provide short descriptions of passive and active treatment technologies that are applicable to the treatment of water from both coal and hard-rock (REE) mining operations. The passive MIW technologies include:

- **Anoxic Limestone Drains (ALDS)**—involve the burial of limestone in oxygen-depleted trenches. MIW is conveyed into these trenches. ALDs generate alkalinity and must be followed by a unit such as an aeration cascade, on or aerobic wetland that oxidizes and removes the precipitated metals. Limestone is a low-cost and effective way to generate alkalinity. However, it must be used in appropriate conditions to ensure its effectiveness. The primary function of ALDs is the treatment of acidic MIW (Spellman, 2017; USEPA, 2014).
- **Successive Alkalinity Producing Systems (SAPS)**—SAPS combine an ALD and a permeable organic substrate into one system that creates anaerobic conditions prior to water contacting the limestone (Skousen, 2000). ASAPS contains a combination of limestone and compost overlain by several feet of water. Mine drainage enters at the top of the pond, flows down through the compost where the drainage gains dissolved organic matter and become more reducing, and then flows into the limestone below, where it gains alkalinity (USEPA, 2004). Dissolution of the limestone raises the pH of the water, resulting in the precipitation of aluminum, copper, and iron. The precipitated metals collect at the base of the SAPS system and in the downstream settling pond.
- **Aluminator©**—is a modified version of Damariscotta's SAPS design. It is an adaptation of a limestone drain in which aluminum hydroxide will accumulate for metals recovery. Instead of using the impervious cap present in limestone drains, the Aluminator© system uses an organic layer and water. The system is designed to operate under high iron and oxygen concentrations, increase pH values and generate alkalinity, retain aluminum in the treatment system, and carry out these functions with a minimum of operation and maintenance requirements (Kepler and McCleary, 1994). The Aluminator© passive treatment system can treat a wide range of MIW flow rates. The Aluminator© treats aluminum in MIW. It can also treat acidity and remove iron. This technology does not appear to have been widely applied.
- **Constructed Wetlands**—are built on the land surface using soil or crushed rock/media and wetland plants. Constructed wetlands can be designed as aerobic wetlands, anaerobic horizontal-flow wetlands, and vertical-flow ponds (vertical-flow wetlands). Constructed treatment wetlands are designed to treat contaminants over a long period and can be used as the sole technology, where appropriate, or as part of a larger treatment approach. Contaminants

are removed through plant uptake, volatilization, and biological reduction. The soil- and water-based microbes remove dissolved and suspended metals from acid mine drainage. The primary advantages of wetland treatment are low capital and operation and maintenance costs. Wetlands can treat acidic, neutral, or alkaline mine drainage.

- **Biochemical Reactors (BCRs or bioreactors)**—use microorganisms to remove contaminants from MIW. BCRs can be constructed in various designs, including in open or buried ponds, in tanks, or in trenches between mine waste and a surface water body (ITRC, 2010). Note that the BCR is not to be confused with the Rotating Biological Contactors (RBSs) used in conventional wastewater treatment operations. The *rotating biological contactor (RBC)* is a biological treatment system (see Figure 10.2) and is a variation of the attached growth idea provided by the trickling filter. Still relying on microorganisms that grow on the surface of a medium, the RBC is instead a **fixed film** biological treatment device—the basic biological process, however, is similar to that occurring in the trickling filter. An RBC consists of a series of closely spaced (mounted side by side), circular, plastic (synthetic) disks, which are typically about 3.5 m in diameter and attached to a rotating horizontal shaft (see Figure 10.2). Approximately 40% of each disk is submersed in a tank containing the wastewater to be treated. As the RBC rotates, the attached biomass film (zoogleal slime) that grows on the surface of the disk move into and out of the wastewater. While submerged in the wastewater, the microorganisms absorb organics; while they are rotated out of the wastewater, they are supplied with needed oxygen for aerobic decomposition. As the zoogleal slime reenters the wastewater, excess solids and waste products are stripped off the media as sloughings. These sloughings are transported with the wastewater flow to a settling tank for removal.

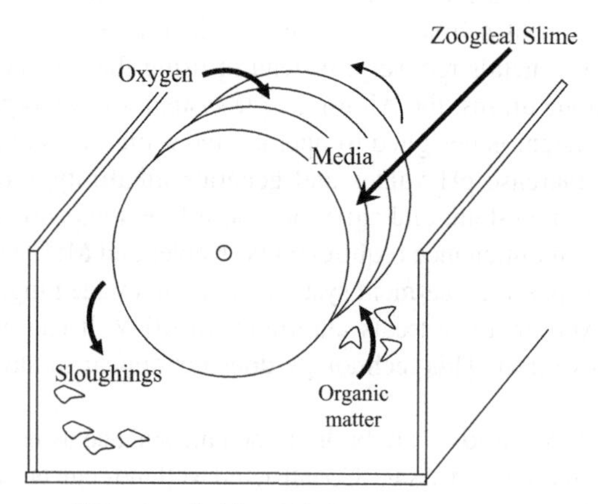

FIGURE 10.2 Rotating biological contactor (RBC) cross-section and treatment system used in conventional wastewater treatment plants.

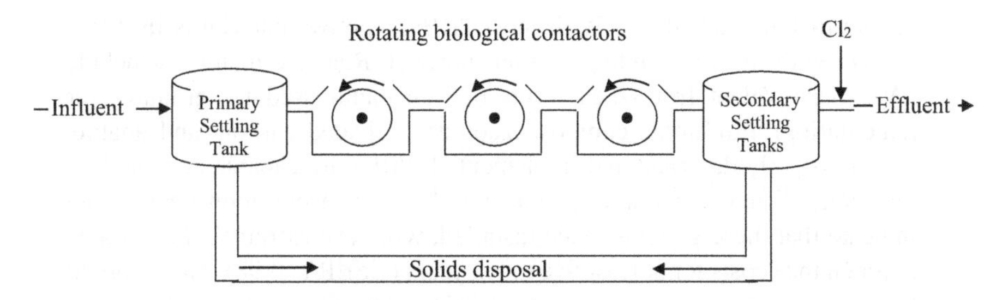

FIGURE 10.3 Rotating biological contactor (RBC) in a conventional wastewater treatment system.

Modular RBC units are placed in series (see Figure 10.3). Simply because a single contactor is not sufficient to achieve the desired level of treatment; the resulting treatment achieved exceeds conventional secondary treatment. Each individual contractor is called a stage and the group is known as a train. Most RBC systems consist of two or more trains with three or more stages in each. The key advantage in using RBCs instead of trickling filters is that RBCs are easier to operate under varying load conditions, since it is easier to keep the solid medium wet at all times. Moreover, the level of nitrification, which can be achieved by a RBC system, is significant—especially when multiple stages are employed.

Again, BCRs operate on the same principle as the RBC systems the main difference is that RBCs treat organic waste and BCRs treat inorganic wastes such as mine waste. Although it must be noted that BCRs can address a wide range of flows, acidity, and metals loading. They can also operate passively or actively. Passive bioreactors have successfully treat acid mine drainage-contaminated waters in pilot and full-scale projects. Most BCRs at mine sites include sulfate-reducing bacteria and operate anaerobically. As the sulfate is reduced in these systems, metal sulfides are precipitated and removed. BCRs effectively increase pH and remove sulfate and metals.

- **Phytotechnologies**—use plants to treat or capture contaminants in various media. Phytoremediation mechanisms include extraction of contaminants from soil or groundwater; hydraulic control of contaminated groundwater; and control of runoff, erosion, and infiltration by vegetative covers. The mechanisms for removal of contamination by Phytotechnologies include concentrations of contaminants in plant tissue; degradation of contaminants by various biotic or abiotic processes; volatilization or transpiration on volatile contaminants form plants to the air; and immobilization of contaminants in the root zone (USEPA, 2000). Identifying the appropriate plant species and soil amendments is essential to treatment success. Long-term maintenance is minimal once the vegetation is established (ITRC, 2010). Phytotechnologies are also used to control run-off, erosion, and infiltration.
- **Permeable Reactive Barriers (PRB)**—consists of a permeable treatment zone in which reactive material has been placed and through which

contaminated water flows. With most PRBs, reactive material is in direct contact with the surrounding aquifer material. Reactive materials include ZVI (Zero Valent Iron—i.e., forms of iron metal used for groundwater remediation), limestone, compost, zeolites, activated carbon, and apatite. When properly designed and implemented, PBRs are capable of remediating many different contaminants to regulatory concentration goals. Data indicate that these systems, once installed, will have extremely low, if any, maintenance costs for at least five to ten years (USEPA, 1998). There should be no operational costs other than routine compliance and performance monitoring. This technology has successfully treated many different constituents, including radionuclides, trace metals, and anion contaminants.

Active MIW technologies include:

- **Fluidized Bed Reactor**—in a fluidized, or pulsed, bed reactor, contaminated water passes through a granular solid media, such as sand or granular activated carbon, at high enough flow rates to fluidize the media, creating a mixed reactor configuration for attached biological growth or biofilm. Fluidized bed reactors (FBRs) can be designed as aerobic or anaerobic systems. Removal of solids following the biological treatment system is required.
- **Reverse Osmosis (RO)**—to gain a fundamental understanding of the RO process it is important to provide foundational information first; the following definitions are provided to facilitate understanding of this process.

    RO Definitions:

    "Miscible," "Solubility"

    1. *Miscible*: capable of being mixed in all proportions. Simply, when two or more substances disperse themselves uniformly in all proportions when brought into contact they are said to be completely soluble in one another, or completely miscible. The precise chemistry definition is: "homogenous molecular dispersion of two or more substances" (Jost, 1992). Examples are:
        - All gases are completely miscible.
        - Water and alcohol are completely miscible.
        - Water and mercury (in its liquid form) are immiscible liquids.
    2. Between the two extremes of miscibility, there is a range of solubility; that is, various substances mix with one another up to a certain proportion. In many environmental situations, a rather small amount of contaminant is soluble in water in contrast to complete miscibility of water and alcohol. The amounts are measured in parts per million (ppm).

    "Suspension," "Sediment," "Particles," "Solids"

    Often water carries solids or particles in suspension. These dispersed particles are much larger than molecules and may be comprised of millions of molecules. The particles may be suspended in flowing conditions and initially under quiescent conditions, but eventually gravity causes the settling of the particles. The resultant accumulation by settling is often called sediment or biosolids (sludge) or residual solids in wastewater treatment vessels. Between this extreme of readily falling out by gravity and permanent dispersal as a solution at the molecular level, there are intermediate types of dispersion or suspension. Particles can be so finely milled or of such small intrinsic size as to remain in suspension almost indefinitely and in some respects similarly to solutions.

"Emulsion"

Emulsions represent a special case of a suspension. As you know, oil and water do not mix. Oil and other hydrocarbons derived from petroleum generally float on water with negligible solubility in water. In many instances, oils may be dispersed as fine oil droplets (an emulsion) in water and not readily separated by floating because of size and/or the addition of dispersal-promoting additives. Oil and, in particular, emulsions can prove detrimental to many treatment technologies and must be treated in the early steps of a multi-step treatment train.

"Ion"

An ion is an electrically charged particle. For example, sodium chloride or table salt forms charged particles on dissolution in water; sodium is positively charged (a cation), and chloride is negatively charged (an anion). Many salts similarly form cations and anions on dissolution in water.

"Mass Concentration"

The concentration of an ion or substance in water is often expressed in terms of parts per million (ppm) or mg/L. Sometimes parts per thousand or parts per trillion (ppt) or parts per billion (ppb) are also used. These are known as units of expression. A ppm is analogous to a full shot glass of swimming pool water as compared to the entire contents of a standard swimming pool full of water. A ppb is analogous to one drop of water from an eye dropper into the total amount of water in a standard swimming pool full of water.

$$\text{ppm} = \frac{\text{mass of substance}}{\text{mass of solutions}} \tag{10.1}$$

Because 1 kg of solution with water as a solvent has a volume of approximately 1 L,

1 ppm ≈ 1 mg/L

"Permeate"

The portion of the feed stream passes through an RO membrane.

"Concentrate" "Reject" "Retentate" "Brine" "Residual Stream"

The membrane output stream that contains water which has not passed through the membrane barrier, and concentrated feedwater constituents that are rejected by the membrane.

"Tonicity"

Is a measure of the effective osmotic pressure gradient (as defined by the water potential of the two solutions) of two solutions separated by a semipermeable membrane? It is important to point out that unlike osmotic pressure, tonicity is influenced only by solutes that cannot cross the membrane, as only these exert an effective osmotic pressure. Solutes able to freely cross do not affect tonicity because they will always be in equal concentrations on both sides of the membrane. There are three classifications of tonicity that one solution can have relative to another. The three are *hypertonic*, *hypotonic*, and *isotonic*.

- **Hypertonic**—refers to a greater concentration. In biology, a hypertonic solution is one with a higher concentration of solutes outside the cell than inside the cell; the cell will lose water by osmosis.
- **Hypotonic**—refers to a lesser concentration. In biology, a hypotonic solution has a lower concentration of solutes outside the cell than inside the cell; the cell will gain water through osmosis.
- **Isotonic**—refers to a solution in which the solute and solvent are equally distributed. In biology, a cell normally wants to remain in an isotonic solution, where the concentration of the liquid inside it equals the concentration of liquid outside it; there will be no net movement of water across the cell membrane.

"Osmosis"

The naturally occurring transport of water through a membrane from a solution of low salt content to a solution of high salt content in order to equalize salt concentrations.

"Osmotic Pressure"

A measurement of the potential energy difference between solutions on either side of a semipermeable membrane due to osmosis. Osmotic pressure is a colligative property, meaning that the property depends on the concentration of the solute, but not on its identity.

"Osmotic Gradient"

The osmotic gradient is the difference in concentration between two solutions on either side of a semipermeable membrane and is used to tell the difference in percentages of the concentration of a specific particle dissolved in a solution. Usually, the osmotic gradient is used while comparing solutions that have a semipermeable membrane between the allowing water to diffuse between the two solutions, toward the hypertonic solutions. Eventually, the force of the column of water on the hypertonic side of the semipermeable membrane will equal the force of diffusion on the hypotonic side, creating equilibrium. When equilibrium is reached, water continues to flow, but it flows both ways in equal amounts as well as force, therefore stabilizing the solution.

"Membrane"

A thin layer of material capable of separating materials as a function of their chemical or physical properties when a driving force is applied.

"Semipermeable Membrane"

A membrane that is permeable only by certain molecules or ions.

"RO System Flow Rating"

Although the influent and reject flows are usually not indicated, the product flow rate is used to rate an RO System. An 600-gpm RO would yield 600 gpm of permeate; thus, it is rated at 600 gpm.

"Recovery" "Conversion"

The ratio of the permeate flow to the feed flow is fixed by the designer and is generally expressed as a percentage. Used to describe what volume percentage of influent water is recovered. Exceeding the design recovery can result in accelerated and increased fouling and scaling of the membranes.

$$\% \text{ Recovery} = (\text{Recovery flow/feed flow}) \times 100 \qquad (10.2)$$

"Concentration Factor" (CF)

The concentration factor is the ratio of solute contamination in the concentrate stream to solute concentration in the feed system. It is related to recovery in that at 40% recovery, the concentrate would be two-fifths that of the influent water.

"Rejection"

The term rejection is used to describe what percentage of an influent species a membrane retains. For example, 97% rejection of salt means that the membrane will retain 97% of the influent salt. It also means that 3% of influent salt will pass through the membrane into the permeate; this is known as salt passage. Equation 10.3 is used to calculate the rejection of a given species.

$$\% \text{ Rejection} = [(C_i - C_p)/C_i] \times 100 \qquad (10.3)$$

where: C_i = influent concentration of a specific component

C_p = permeate concentration of a specific component

RO constituents and processes provide a proven method to demineralize acid mine drainage. However, it does require significant construction and operating costs.

- **Zero Valent Iron (ZVI)**—can be used in active MIW treatment systems to rapidly neutralize acid and promote the removal and immobilization of dissolved heavy metals. Adsorption onto the iron metal surface, or onto iron corrosion products initially present on the unreacted metal surface, facilities the rapid remove of metals from MIW. ZVI can reduce selenium oxyanions to elemental selenium. Ferrous cations can also reduce selenate to selenite and subsequently remove selenite by adsorption to iron hydroxides. In an aqueous environment, ZVI can be oxidized to dissolved ferric (Fe^{3+}) and ferrous (Fe^{2+}) ions. The ions react with hydroxyl ions present in water to form ferric and ferrous hydroxides. Selenate is reduced selenite while ferrous iron is oxidized to ferric iron. Selenite then adsorbs to the ferric and ferrous hydroxide surfaces and is removed from the solution.

- **Rotating Cylinder Treatment Systems™ (RCTS)**—is an innovative form of lime precipitation treatment. Lime is used to increase the pH of the contaminated waste, allowing of the oxidation or precipitation of metals (ITRC, 2010). In typical lime treatment systems, scaling of precipitated metal hydroxides and oxides can reduce or eliminate the efficiency of the system. In an RCTS, water being treated flows through shallow troughs containing rotating perforated cylinders that transfer oxygen and agitate the water, intending to reduce or prevent scaling. RCTS systems can treat highly acidic waters, high sulfate, and metals in cold weather or remoted locations. RCTS require a post-treatment solids separation unit, such as a pond. RCTS systems precipitate metals and increase pH.

  In 2004 and 2005, two four-rotor RCTS units were installed in series to treat acid mine drainage containing more than 7 grams per liter (g/L) of dissolved metals at the Rio Tinto Mine (see Figure 10.4) (Tsukamoto, 2006). Lime slurry was mixed with impacted water in a lime-dosing tank to raise the pH from about 2.6 to about 8.5. the water from the flash rector tank was then mixed with RCTS effluent prior to delivery to a settling pond. The pH-adjusted mixture from the settling pond was then fed into the RCTS system. The flow rate to the system ranged from 5 to 20 gpm during two treatment events. The RCTS treatment system effectively removed metals to below applicable water standards at the site. Dissolved metal concentrations were reduced to less than 0.2 mg/L, with the exception of manganese, which was reduced from up to 96 mg/L to less than 0.58 mg/L.

  The RCTS has been shown to be effective at treating acid mine drainage, sulfate, and neutral or alkaline drainage. A 2008 demonstration conducted near Gladstone, Colorado, suggested that the system is capable of meeting the applicable water quality criteria for the constituents of concern (Smart et al., 2009).

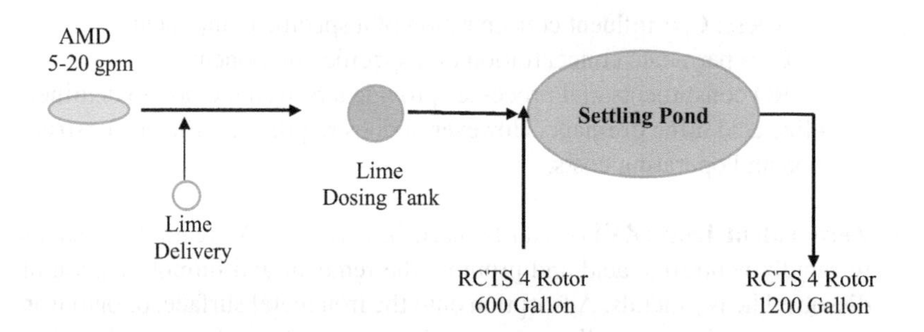

FIGURE 10.4 Schematic of the 2005 Rio Tinton, Nevada Water Treatment system.

- **Ferrihydrite Adsorption (aka co-precipitation)**—is a two-step physical adsorption process that can remove heavy metals from MIW. In low-iron-containing waters, iron may be added to co-precipitate or adsorb certain metals onto ferric hydroxide precipitates. The process involves the addition of a ferric salt to the water to generate a ferric hydroxide and ferrihydrite precipitate, the formation of which results in concurrent adsorption of metals on the surface. The precipitated iron can then be removed. EPA designated the technology a best Demonstrated Available Technology for selenium removal. It is widely implemented at a full scale throughout the mining industry. Constituents commonly treated by this technology include metals that can co-precipitate with iron.
- **Electrocoagulation**—involves the application of an electrical current to coagulate organic constituents and suspended solids in water. The affected water is treated using electrolysis with graphite or stainless-steel cathodes in conjunction with a metal anode (Golder, 2009). When a voltage is applied, metals precipitate out of the water. A secondary treatment step such as sedimentation of filtration can then remove the precipitated metals from the water. Mining operators use electrocoagulation to remove suspended particles of clay and coal fines from mine process water. However, electrocoagulation is not a proven technology for full-scale treatment of mining wastes.
- **Ion Exchange**—in standard water/wastewater treatment, an ion exchange softener is a common alternative to the use of lime and soda ash for softening water. Natural water sources contain dissolved minerals that dissociate in water to form charged particles called *ions*. Of main concern are the positively charged ions of calcium, magnesium, and sodium, and bicarbonate, sulfate, and chloride are the normal negatively charged ions of concern. An ion exchange medium, called *resin*, is a material that will exchange a hardness-causing ion for another one that does not cause hardness, hold the new ion temporarily, and then release it when a regenerating solution is poured over the resin. The removal capacity of an exchange resin is generally reported as grains of hardness removal per cubic foot of resin. To calculate the removal capacity of the softener, we use Equation (10.4).

$$\text{Exchange Capacity, grains} = (\text{Removal Cap., grains/ft}^3)\,(\text{Media Vol., ft}^3) \quad (10.4)$$

Example 10.1

The hardness removal capacity of an exchange resin is 24,000 grains/ft³. If the softener contains a total of 70 ft³ of resin, what is the total exchange capacity (grains) of the softener?

$$\begin{aligned}\text{Exchange Cap., grains} &= (\text{Removal Cap., grains/ft}^3)\,(\text{Media Vol., ft}^3)\\ &=(22{,}000\ \text{grains/ft}^3)\,(70\ \text{ft}^3)\\ &=1{,}540{,}000\ \text{grains}\end{aligned}$$

Example 10.2

Problem: An ion exchange water softener has a diameter of 7 ft. The depth of resin is 5 ft. If the resin has a removal capacity of 22 kg/ft^3, what is the total exchange capacity of the softener (in grains)?

Solution: Before the exchange capacity of a softener can be calculated, the ft^3 resin volume must be known:

$$\text{Vol., ft}^3 = (0.785)\ (D^2)\ (\text{Depth, ft})$$
$$= (0.785)\ (7\text{ ft})\ (7\text{ ft})\ (5\text{ ft})$$
$$= 192\text{ ft}^3$$

Calculate the exchange capacity of the softener:

$$\text{Exchange Cap., grains} = (\text{Removal Cap., grains/ft})\ (\text{Media Vol., ft}^3)$$
$$= (22{,}000\text{ grains/ft}^3)\ (192\text{ ft}^3)$$
$$= 4{,}224{,}000\text{ grains}$$

Hardness can be removed by ion exchange. In water softening, ion exchange replaces calcium and magnesium with a non-hardness cation, usually sodium. Calcium and magnesium in solution are removed by interchange with sodium within a solids interface (matrix) through which the flow is passed. Similar to the filter, the ion exchanger contains a bed of granular material, a flow distributor, and an effluent vessel that collects the product. The exchange media include greensand (a sand or sediment given a dark greenish color by grains of glauconite), aluminum silicates, synthetic siliceous gels, bentonite clay, sulfonated coal, and synthetic organic resins and are generally in particle form usually ranging up to a diameter of 0.5 mm. Modern applications more often employ artificial organic resins. These clear, BB-sized resins are sphere-shaped and have the advantage of providing a greater number of exchange sites. Each of these resin spheres contains sodium ions, which are released into the water in exchange for calcium and magnesium. As long as exchange sites are available, the reaction is virtually instantaneous and complete.

When all the exchange sites have been utilized, hardness begins to appear in the influent (breakthrough). When breakthrough occurs, this necessitates the regeneration of the medium by contacting it with a concentrated solution of sodium chloride.

Ion exchange used in water softening has both advantages and disadvantages. One of its major advantages is that it produces a softer water than chemical precipitation. Additionally, ion exchange does not produce the large quantity of sludge encountered in the lime-soda process. One disadvantage is that, although it does not produce sludge, ion exchange does produce concentrated brine. Moreover, the water must be free of turbidity and particulate matter or the resin might function as a filter and become plugged.

In mine wastewater treatment, ion exchange still the same principle whereby the reversible exchange of contaminant ions with more desirable ions of a similar charge is adsorbed to solid surfaces known as ion exchange resins. The active process provides hardness removal, desalination, alkalinity removal, radioactive waste removal, ammonia removal, and metals removal. Depending on the type of water is to be treated, selective metal recovery may be an option.

- **Biological Reduction**—is a process to remove elevated levels of selenium, metals, and nitrate found in wastewater streams near mining operations, coal-fired power plants, and other industries. This system falls broadly into the category of active BCR systems. The system is a low-energy system and uses biofilters seeded with selected strains of naturally occurring nonpathogenic microorganisms to produce treated effluent that meets or exceeds regulatory standards for selenium removal. The microorganisms in anaerobic bioreactors reduce selenium in the form of selenite and selenate to elemental selenium. The end product is a fine precipitate of elemental selenium that is removed from the bioreactor during period backflushing. Several pilot—scale tests have proven the reliability of this process. It is now operational full-scale at several locations worldwide. Biological reduction removes elevated levels metals in wastewater.
- **Ceramic Microfiltration**—is a pressure-driven membrane separation process designed to remove heavy metals from acid mine drainage. It uses ceramic membrane microfilters to remove precipitated solids that allow for effective cleaning to restore membrane permeability. Ceramic microfiltration requires significant energy input and therefore has high operation costs. However, it may be more cost-effective than other alternatives since it may replace several unit treatment processes with a single process (USEPA, 2012).

Reuse of mine wastewater is a common practice. The wastewater is removed from underground workings or open pits or ponds and then re-circulated for onsite use (e.g., processing areas) or even offsite use if the water is of adequate quality. However, these waters must meet discharge limits for concentrations of metals or other contaminants and also be permitted for discharge to surface waters. The treatment technologies discussed in this chapter can work to make the mining wastewater safe for reuse.

Mine Soil Storage Piles

In order to mine an aboveground mine soil must be removed. Upon removal, the soil must be placed somewhere and the general practice is to place the removed soil in storage piles. These storage piles continuously grow in size as more soil from the mining operations is removed. Typically, the soil storage pile is reused whenever the mine is restored at the end of the mine's life. Although soils are not generally considered wastes if they are not managed properly, especially if they become a source of sediment runoff. The mine soils removed in the mining process may be humus-rick and acidic, and fines that run off to streams could affect surface water quality. Moreover, acidic soils used as cover for metal-bearing rock-waste area could encourage metal leaching from the rock.

Overburden

The material removed to expose the ore of interest during mining is called overburden. Overburden storage piles of soil, sediments, and rock materials that do not contain the commodity can include other metals and constituents that may also be a source of pollution and contamination. While relatively small in comparison to quantities of other mining waste overburden may include unconsolidated materials such as alluvium, colluviums (i.e., materials that accumulate at the foot of a steep slope), glacial deposits, or residuum (i.e., substances such as chemicals left behind) and be a source of sediment or acid-generating minerals; however, the acidity of these materials is usually low and they generally do not contain significant concentrations of metals. Overburden usually does not include rock that contains the mineral ore, but it may include the very low mineralized rock removed from around the ore from hard rock mines. These materials can contain greater metal concentrations than typically associated with unconsolidated deposits. Without proper management, erosion of the overburden storage piles resulting in sediment loading to steams and surface water can occur. Groundwater may be entrained with the overburden material, which can then leach through to the base of the pile, carrying with it high concentrations of metals and other potentially harmful constituents. Overburden piles can cover hundreds of acres and be a prominent feature in the general landscape as they reach heights extending hundreds of feet. Large mines may have multiple overburden storage piles. Large mine trucks and other vehicles deliver materials to the top of the pile and shape the slopes. Fugitive dusts are a concern from overburdened storage piles.

Ore Stockpiles

Ores may be stockpiled to provide a supply, lasting six months or more, of ore stock feed for beneficiation (ore dressing) and processing plants. Subeconomic ore is often stockpiled at the mine site for future exploitation under the appropriate economic or market demand conditions. These ore stockpiles and subeconomic ore storage piles can include large areas within multiple sites. These ore stockpiles can represent a significant source of toxic metals. The nearby aquatic habitat can suffer significant effects from runoff, flooding, or infiltration, or rainwater if not captured and managed properly. Groundwater may be entrained with the ore material, which can from leach out to the base of the pile, carrying with it high concentrations of metals and other potentially harmful constituents. Large mine trucks and other vehicles deliver materials to the top of the pile and shape the slopes. Fugitive dusts containing metals or other harmful constituents form the ore and subeconomic ore stockpiles, even though not as much of a concern as tailings wastes of conveyances, may still be of potential concern depending upon the material stored (USEPA, 2012).

As previously discussed, acid mine waste (AMD) is uncommon in most REE deposits; however, some potential exists for low levels of acid generation from accessory minor sulfide minerals, especially in low-Ti iron oxide Cu–U–Au–REE deposits like Pea Ridge (Foose and Graunch, 1995). The carbonatite ores at the Mountain Pass mine, for example, may cause elevated pH levels in streams, depending on the concentration of other acid-producing accessory minerals in the ore-bearing rock and surrounding country rock that can serve to neutralize any runoff. Thorium-rare earth element veins have moderately high sulfur content, but the sulfur is present chiefly as sulfate in barite. The sulfide mineral content of these vein deposits is very low, thus potential for AMD generation is low (Armbrustmacher et al., 1995). Carbonatites are important REE ores, and these igneous rocks contain greater than 50% carbonate minerals. Waste rock from REE deposits could potentially present a problem with neutral mine drainage (NMD), with pH in the range of 6 to 10. Mine drainage in the NMD pH range can have various elevated metals (e.g., zinc, cadmium, manganese, antimony, arsenic, and selenium) (INAP, 2010). In the case of REE deposits, elements like uranium and vanadium could be mobile under NMD conditions, and these elements are constituents of some REE ores. Particular engineering designs, waste handling and disposal procedures, or closure and reclamation plans may be reused for those materials whose characteristics may pose significant risks. The impact depends upon the ability of receiving wastes (groundwater and surface water) to neutralize and dilute the drainage.

Waste Rock

Note that the following information about mining waste rock is adapted from EPA and *Hardrock Mining: A Source Book for Industry in the Northwest and Alaska*

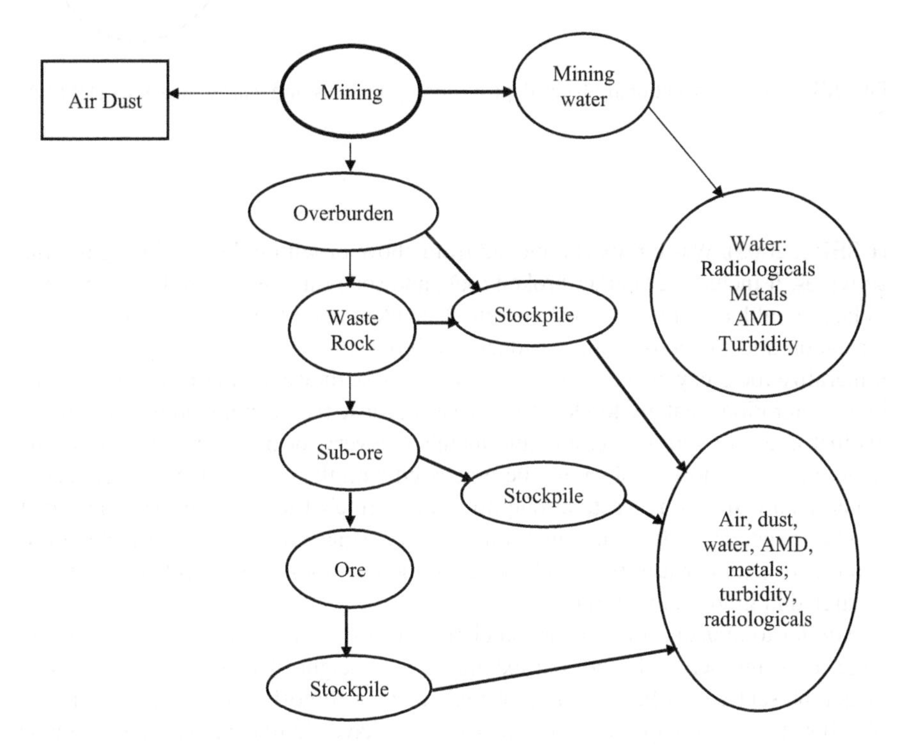

FIGURE 10.5 Conventional hard rock deposit mining process and waste emissions (USEPA, 2012).

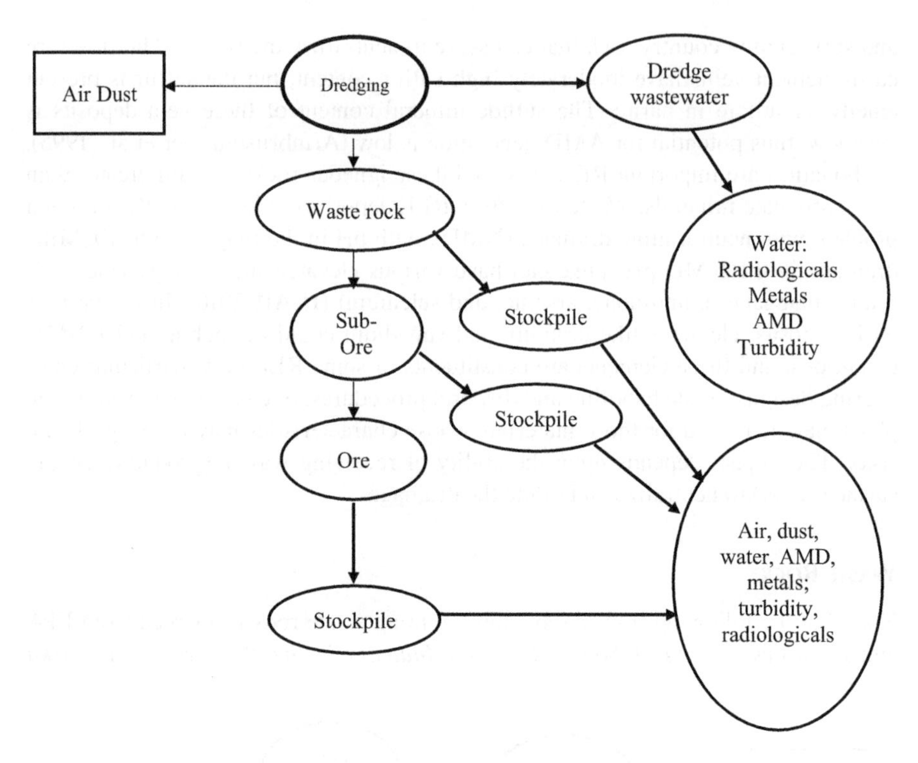

FIGURE 10.6 Conventional placer deposit mining process and waste emissions (USEPA, 2012).

(USEPA, 2003). Waste rock is removed from above or within the ore during mining activities. It includes granular, broken rock, and soils ranging in size from fine sand to large boulders, with the fines' content dependent upon the nature of the geologic formation and methods employed during mining. Non-mineralized and low-grade mineralize rock may be designated as waste because they contain the target minerals in concentrations that are too low to process, because they contain additional minerals that interfere with processing and metals recovery, or because they contain the target metal in a form that cannot be processed with the existing technology. These materials are stored as waste at one point in a mine's life, but may become ore at another stage, depending on commodity prices, changes in and costs of technology, and other factors. Waste rock and subeconomic ores may be stockpiled together or separated in grades of material.

Similar to ore/subeconomic ore stockpiles, waste rock storage piles are typically large, covering acres of land and extending to a height of many feet. Waste rock piles represent a significant source of toxic metals. Runoff, flooding, or infiltration of rainwater, if not captured and managed, can have significant effects on aquatic habitats. Groundwater may be entrained with the waste rock if the rock units being mined occur below the water table, which can then leach out to the base of the pile, carrying with its concentrations of metals and other potentially harmful constituents. Large mine trucks and other vehicles deliver materials to the top of the pile and shape the slopes. Fugitive dusts containing metals common to the ore material are also a concern from waste rock storage piles. Figures 10.5 and 10.6 present block flow diagrams of conventional hard rock and placer deposit mining, materials management, and potential pollutants.

REFERENCES

Armbrustmacher, T.J., Modreski, P.J., et al. (1995). Mineral deposits models: the rare earth element vein deposits. United State Geological Survey. Accessed 12/28/2021 @ http://www.pubs.usgs.gov/of/1995/ofr-95-0831/CHAP7.pdf.

Betournay, M.C. (2011). Underground mining and its surface effects. Accessed 12/23/2021 @ http://www.fhwas.dot.gov.engineering/geotech/hazards/mine/workshops/iawkshp/betourna2.cfm.

Foose, M.P. and Graunch, V.J.S. (1995). Low Ti iron oxide Cu-U-Au-REE deposits (Models 25i and 29b; Cox. 1986a, b: summary of relevant geologic, geoenvironmental and geophysical information. United States Geological Survey Open-file Report No. 95–0831 (Chapter 22). Washington, DC: USGS.

Golder Associates, Inc. (2009). Literature review of treatment technologies to remove selenium from mining influenced water. Accessed 01/01/2022 @ http://name.org/docs00057713.PDF.

Gruber, P.W., Medina, P.A., Keoleian, G.A., Kesler, S.E., Everson, M.P. and Washington, T.J. (2011). Global lithium availability: a constraint for electric vehicles? *Journal of Industrial Ecology*. Accessed 12/23/2021 @ http://www.eenews.net/assets/2011/07/27/document _gw_02.pdf.

International Atomic Energy Agency (2005). Guidebook on environmental impact assessment for in situ leach mining projects. Accessed 12/23/2021 @ https://www-pub.iaea.org/MTCD/publications/PDF/TE_1428_web.pdf.

International Network for Acid Prevention (2010). Global Acid Rock Drainage Guide (GUARD Guide), Version 0.8. Accessed 01/01/2022 @ http://gardguide.com/index.php?title=Main_Page.

Interstate Technology Regulatory Council (2010). Mining waste treatment technology selection website. Accessed 12/31/2021 @ http://www.itrcweb.org/miningwaste-guidance/.

Jost, J. (1992). Homogenous molecule dispersion. Accessed 12/26/ 21 @ https://www.osti.gov/biblio/4243132.

Kepler, D. and McCleary, E. (1994). Successive alkalinity-producing systems (SAPS) for the treatment acidic mine drainage. Proceedings of the International Land Reclamation and Mine Drainage Conference, April 24–29, 1994, USDI, Bureau of Mines SP 06A-94, Pittsburg, PA, 195–204.

Reichardt, C. (2008). Heap leaching and the water environment.—Does low-cost recovery come at a high environmental cost? Accessed 12/23/21 @ http://www.imwa.info/docs.imwa_2008.

Skousen, J. (2000). Overview of passive systems for treating acid mine drainage. Reclamation of drastically disturbed lands. (Agronomy, No. 41). Accessed 01/01/2022 @ http://www.wvu.edu/-agexten/landrec/passtrt/passtrt.htm.

Smart, P., Reisman, D., Odell, S., Forrest, S., Ford, K. and Tsukamoto, T. (2009). Rotating cylinder treatment system demonstration. CO. Proceedings, 26th Annual American Society of Mining and Reclamation Conference, May 30–June 5, billings, MT.

Spellman, F.R. (2017). *The Science of Environmental Pollution*, 3rd ed. Boca Raton, FL: CRC Press.

Tsukamoto, T. (2006). High efficiency modular treatment of acid mine drainage field applications at Wester U.S. Sites with the Rotating Cylinder Treatment System (RCTS). Accessed 01/01/2022 @ http://www.iwtechnolgies.com/pdfs/High_Efficiency_Treatment_Acid_Mine_Drainage.pdf.

USEPA (1998). Permeable reactive barriers. Technologies for reclamation. Accessed 01/01/2022 @ http://clu-in.org/download/rtdf.prb/reactar.pdf.

USEPA (2000). Mine waste technology annual report. Washington, DC: United States Environmental Protection Agency.

USEPA (2003). EPA and hardrock mining: a source book for industry in the Northwest and Alaska: Appendix G: aquatic resources. Washington, DC: United States Environmental Protection Agency.

USEPA (2004). Demonstration of Aquafix and SAPS Passive Mine Water Treatment Technologies at the Summerville Mine Site. EPA/540/R-04/501. Accessed 01/01/2022 @ http:www.epa.gov/aml/tech/news/summityl.htm.

USEPA (2012). Rare earth elements: a review of production, processing, recycling, and associated environmental issues. EPA 600/R-12/572. Accessed 12/23/2021 @ www.epa.gov/ord.

USEPA (2014). Reference guide to treatment technologies for mining-influenced water. EPA 542-R-14-001. Washington, DC: United Stated Environmental Protection Agency.

World Nuclear Association (2012). In situ leach (ISL) mining of uranium. World Nuclear Association. Accessed 12/23/2021 @ http://www.world-nuclear.org/info/inf27.html.

11 REE Processing

When describing a process, any process, that is complex and has a high potential for environmental contamination, for leaving a very ugly footprint, and is poorly monitored, poorly controlled, and poorly managed one need go no further than the milling or rare earth elements. Waste streams from REE processing have been identified, and their hazardous waste potential assessed. It might surprise you to know that the REE liquid waste streams even though they can be contaminators of the environment are not the worst contributors to pollution; instead, it is the tailings and associated treatment and storage that possess the greatest pollution potential. The simple truth is that when not controlled heavy metals and radionuclides associated with REE tailings pose the greatest threat to the environment and human health. The good news is that new technologies and management practices are showing the potential to reduce the environmental risk of environmental contamination (USEPA, 2012).

PROCESSING

As mentioned, the two major sources of REEs are bastnasite and monazite and are the focus of this chapter. The mineral bastnasite is a member of the carbonate–fluoride family. Most bastnasite is bastnasite-(CE), and cerium is by far the most common of the rare earth in this particular class of mineral. Bastnasite and the phosphate mineral monazite are the two largest sources of cerium and other rare-earth elements. However, it is important to note that production can come from a variety of minerals, such as xenotime, apatite, yttrofluorite, cerite, and gadolinite. Because of their strong affinity for oxygen, REEs are primarily present as oxidic compounds, and resources are often expressed as rare-earths oxides (REOs). Processing REOs into usable products is a very multifaceted procedure and often differs considerably between deposits. The selection of different procedures used in treatment processes is affected by the major factors listed below (Ferron et al., 1991):

- Type and nature of the deposit (e.g., beach sand, vein type, igneous and complex ores) and its complexity
- Type and nature of other valuable minerals present with REOs
- Type and nature of gangue minerals present in the deposit (e.g., slimes, clay, and soluble gangue)
- Type and composition of the individual REO minerals
- The social and environmental acceptability of the process.

The primary steps involved in processing REOs are separation and concentration from the host material in acidic or alkaline solutions, separation of the REOs using solvent extraction of ion exchange, and reduction of the individual REOs into pure metals (Gupta and Krishnamurthy, 2004; Tran, 1991). The first step mine the

DOI: 10.1201/9781003350811-13

hardrock deposit or to dredge the placer deposit. Then crushing the ore and separating the REO by flotation, magnetic, or gravimetric separation. This separation procedure/process significantly increases the percentage of REOs in the working material. A good example of this is the Mountain Pass mine which separates REOs in bastnasite from 7% to 60%. The problem? The problem is waste. A tremendous amount of discarded waste rock (tailings) is generated in this process and is typically managed onsite or used as backfill material. In this first step, chemical changes usually do not occur and this process is usually located near the mine site to reduce transport costs that can be quite extensive.

Subsequent steps in the process intend to change the concentrated mineral into more valuable chemical forms through various thermal and chemical reactions. Typically utilizing leaching, extraction, and precipitation (hydrometallurgy techniques), the mineral concentrates are separated into usable oxides. In order to refine the oxides or metal mixtures into high-purity REMs further processing by metallothermic reduction can be utilized.

Figure 11.1 presents a simple block diagram of typical extractions and processing steps involved in placer mining operations. The simplicity of this diagram is obvious, but what is not so obvious is the potential offshoot waste emissions that many of the steps shown produce. For example, in dredging processing air dust and wastewater containing radiologicals, metals, and turbidity is produced; waste rock is produced and stockpiled; sub-ore is produced and stockpiled; and ore is produced and stockpiled. During grinding and acid digestion, and filtration air dust is produced. In the magnetic separation, process tailings are produced. The tailings are impounded and liquid waste is present. All of these steps have environmental impacts.

Figure 11.2 presents a simple conventional hardrock processing operation for REE with basically the same potential environmental impacts.

Mineral Wastes: Bevill Amendment

In 1980, Congress amended the Resource Conservation and Recovery Act (RCRA) to temporarily exclude wastes from extraction, crushing and sizing (beneficiation), and processing ores and minerals from hazardous waste regulation. Known as the Bevill Amendment, this exclusion was temporary until Environmental Protection Agency (EPA) completed a Report to Congress and made a regulatory determination on these wastes and reported this information in the final rule (54 Fed. Reg.36592, 36616 codified at 261.4(b)(7)).

Later, in 1985, EPA proposed to narrow its 1980 interpretation of the Bevill Amendment for mineral processing wastes. However, EPA later withdrew this proposal because it believed the proposal was inadequate. This withdrawal led to a lawsuit by the Environmental Defense Fund and the Hazardous Waste Treatment Council, who claimed that the Bevill exclusion should include only so-called "special wastes: (i.e., only mineral processing wastes that are both "high-volume" and "low-hazard" wastes). In July 1988, the court determined that six smelting wastes were clearly hazardous and ordered EPA to list them as hazardous. EPA did so on August 31, 1988.

The EPA determined the line between beneficiation and processing of rare earths for one particular rare earth operation is when the ore is digested with concentrated

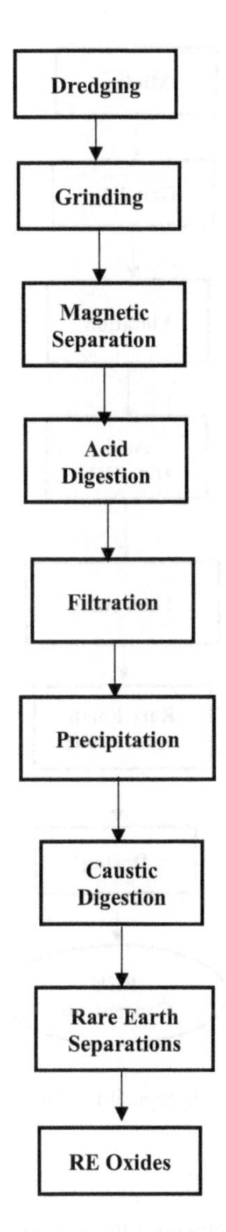

FIGURE 11.1 Conventional placer deposit resource processing.

acids or caustics (USEPA, 1991). However, several factors are involved in making Bevill determinations, and official assistance should be sought from the RCRA authorized state or the EPA Regional office. The basic steps in making Bevill determinations are the following (USEPA, 2012):

1. Determine whether the material is considered a solid waste under RCRA.
2. Determine whether the facility is using a primary ore or mineral to produce a final or intermediate product, and also whether less than 50% of the feedstock on an annual basis are from secondary sources.

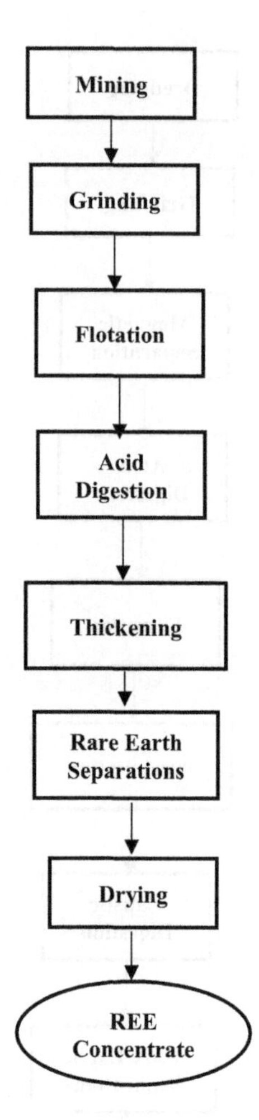

FIGURE 11.2 Conventional hardrock deposit resource processing for REE.

3. Establish whether the material and the operation that generates it are uniquely associated with mineral production.
4. Determine where in the sequence of operations beneficiation ends and mineral processing begins.
5. If the material is a mineral processing waste, determine whether it is one of the 20 special wastes from mineral processing.

This process of determination will result in one of the three outcomes:

- The material is not a solid waste and therefore not subject to RCRA
- The material is a solid waste, but is exempt from RCRA Subtitle C because of the Mining Waste Exclusion or

- The material is a solid waste that is not exempt from RCRA Subtitle C and is subject to regulation as a hazardous waste if it is listed or characteristic hazardous waste.

Beneficiation Processes

Simply stated the beneficiation process is the transformation of a mineral (or a combination of minerals) to a higher-value product that can either be consumed locally or exported. The process does not alter the chemical composition of the ore; rather, these processes are intended to liberate the mineral ore from the host material. The crushing and grinding steps enable the exploitation of differences in physical properties such as gravimetric, magnetizability, and surface ionization potential to aid in separation (Aplan, 1988). The separation processes typically employed are (1) gravity separators, (2) electrical/magnetic separators, and (3) flotation separators.

By exploiting the variance in their gravity-driven movement through a viscous fluid gravity concentration methods separate minerals of different specific gravity. For successful separation using this technique, there must exist a marked density difference between the mineral and the gangue. Examples of gravity separators include (1) jigs; (2) pinched sluices and cones; (3) spirals; (4) shaking tables; (5) pneumatic tables; (6) duplex concentrators; (7) Mozley laboratory separator; and (8) centrifugal concentrators.

Note that electrical and magnetic separation methods are considered similar because they exploit some of the same properties of the mineral and thus their application can overlap. Magnetic separators utilize the differences in magnetic properties between the mineral of interest and gangue. There are two primary classifications of materials when applying magnetic separation: (1) diamagnetic and (2) paramagnetic. Diamagnetic materials are repelled along the lines of magnetic force, while paramagnetic materials are attracted along the lines of magnetic force. There are two primary classes of magnetic separators: low- and high-intensity process units. Highly paramagnetic materials like ferro-magnetic metals are typically separated using low-intensity separators, such as drum separators. Specifically for rare earths, roll separators that utilize alternate magnetic and non-magnetic laminations are commonly used. To separate very weak paramagnetic minerals and include common configurations such as induced roll magnetic separators and Jones separators high-intensity separators are used. Electrical separation exploits electrical conductivity differences among various minerals in the ore. To ensure optimum results, the ore feed needs to be completely dry and only one particle deep. This restriction has limited the application primarily to beach and stream placers, such as those where REEs are found. Plate and screen electrostatic separators are examples of electrical separators.

Floatation is typically used in applications where the ore particle size is too small for efficient gravity separation. Flotation takes advantage of the hydrophobicity (i.e., repels or fails to mix with water) of the mineral or interest and the hydrophilicity (i.e., mixes with water) of the gangue. Accordingly, the hydrophobic mineral particles tend to "stick" to the air bubbles that are delivered into the process unit and rise to the surface where they are separated. A variety of chemicals are added to the ore slurry to enhance the process including collectors, frothers, and modifiers.

Bastnasite Beneficiation

There are several variations in the beneficiation of bastnasite but the process generally involves crushing/grinding and separation by flotation. The best way in which to provide a relevant example is to present in the following a real-world illustration of the details involved in the operation of the Molycorp Mountain Pass mine. The ore containing bastnasite (7% rare earth oxide) is crushed, ground, and classified in the milling process to achieve 100% passing a 150-mesh sieve prior to separation by hot froth flotation (Gupta and Krishnamurthy, 2004). Note that prior to flotation, the ore passes through six different conditioning treatment sin which steam, soda ash, sodium fluosilicate, sodium lignosulfonate, and steam-distilled tall oil (i.e., a by-product of the kraft process of wood pulp manufacture) are added to aid the separation of the unwanted materials (often referred to as gangue). This process produces a 60% REO bastnasite concentrate.

Monazite/Xenotime Beneficiation

Monazite or xenotime ore, typically associated with dredged mineral sands, is separated and concentrated after coarse grinding via gravimetric, flotation, or magnetic processes. As expected, the complexity of this process is dependent on the specific reserve.

Extraction Processes

The most common chemical extraction method of separating individual REOs from the mineral concentrates hydrometallurgy. Note that it is basicity differences between the various rare earths that include the solubility of their salts, the hydrolysis of ions, and the formation of complex specifies (Gupta and Krishnamurthy, 2004). The differences in these properties are exploited by fractional crystallization, fractional precipitation, ion exchange, and solvent extraction to separate the individual REOs. It is important to point out that although some of the individual REOs and rare earth chlorides resulting from these processes have market value, further processing and refining are required to produce high-quality pure metal end products to maximize value. These processes are also utilized to recover REEs from recycled materials. Table 11.1 presents a list of rare earth extraction methods and a brief description of each.

Bastnasite Extraction

The desired end product(s) dictates the subsequent processing steps of the bastnasite concentrate. Typical processes include leaching, washing, filtering, and drying or calcining to increase the percent REO from approximately 60% up to as much as 90%. In the previous operation, the Mountain Pass Mine produced three commercial grades of bastnasite, with end uses such as glass polishing powders and primary alloys for iron and steel production. To produce individual lanthanides, the concentrate was first calcined to convert the contained cerium to plus four valency while leaving the other lanthanides in the plus three valency. Acid digestion followed and resulted in the dissolution of most of the non-cerium lanthanides. The

TABLE 11.1
Rare Earth Extraction Methods (adapted from Meyer and Bras, 2011)

Method	Type	Extraction Trait	Process
Liquid–liquid extraction	Hydrometallurgy	Solubility	The liquid containing the desired element is mixed with an immiscible solvent, which preferentially dissolves the desired element. When the liquids separate, the desired element separates with the solvent.
Solid–liquid	Hydrometallurgy	Solubility	The solid is placed into a solvent, which dissolves the desired soluble component.
Solid phase	Hydrometallurgy	Solubility	The fluid containing the desired element is poured through a sorbent bed, which forms equilibrium by the liquid adsorption to the solid surface or penetration of the out layer of the molecules on that source. Either undesired components can be washed-out or elutriants can be used to selectively extract the desired elements.
Ion exchange	Hydrometallurgy	Chemical affinity	The fluid containing the desired elements is mixed with an elutriant and poured through a resin. The molecules are separated based on their affinity split between the elutriant and the resin.
Super critical	Hydrometallurgy	Variety	The fluid containing the desired element undergoes a reaction with CO_2 at or over the critical temperature of 31°C and critical pressure of 72.9 atm. This amplifies minute differences between elements to allow separation.
Electrowinning	Electrometallurgy	Electronegativity	A current is passed from an inert anode through a liquid bleach solution containing the metal. The metal is extracted by an electroplating process, which deposits the rare earths onto the cathode.
Electrorefining	Electrometallurgy	Electronegativity	The anode is composed of the recycled material. When the current passes form the anode to the cathode through the acidic electrolyte, the anode corrodes, releasing the rare earth solution into the solution, then electrowinning occurs.
Electro slag	Pyrometallurgy	Density	Electricity melts the metal. The molten metal is combined with a reactive flux, which causes the impurities to float off the molten metal into the slag.

Note: With regard to Mountain Pass Mine (Molycorp Metals and Alloys) that owned the Mountain Pass rare earth mine in 2015 it filed for bankruptcy for several reasons. It was purchased by its largest creditor Oaktree Capital Management and was reorganized as Neo Performance Materials. For the purpose of this presentation, the procedures and processes used in rare earth mining by Molycorp are discussed and described namely because the procedures/processes used remain the same with only minor adjustments and most of these are in marketing and other economic areas of concern.

resulting solution was then processed using multistage solvent extraction to produce high-purity rare earth compounds, such as the following:

- neodymium-praseodymium carbonate
- lanthanum hydrate
- cerium concentrate
- samarium oxide
- gadolinium oxide
- terbium oxide
- europium oxide

Monazite/Xenotime Extraction

A common method of processing monazite/xenotimes is taking the ore concentrate from the beneficiation process and digesting it using 70% sodium hydroxide (aka caustic—NaOH) to produce rare earth hydroxides. The rare earth oxides are then leached with HCl to recover the soluble rare earth chlorides. The rare earth chloride solution is then processed using multistage solvent extraction to produce individual, high-purity REOs (95–99.995%).

Tailings Extraction

The tailings waste from other mineral processing operations are not to be ignored for they are another potential source of REO. This is best seen in apatite tailings from the Pea Ridge iron ore mine and processing operation where 0.5% contain REO. The two primary methods for extracting REO from apatite are selective acid extraction and physical separation techniques. The United States Bureau of Mines reports that using gravimetric processes it was able to recover 90% of the REO and produce a 70% lanthanide concentrate (US DOI, 2010). In detail, the pulped apatite was mixed with oleic acid to collect the phosphate when pine oil was used as the frother. Three successive flotation steps yielded the desired lanthanides. The tailings impoundment current contains 20 million tones which is over 7% apatite.

Liberation of REMs

The liberation of REM from compounds such as oxides or chlorides is accomplished by reduction processes. The liberation process via reduction, however, can be a very difficult process due to the oxides extreme stability. Several methods have been developed to accomplish this task. However, the three primary methods of producing REMs are (1) reduction of anhydrous chlorides or fluorides, (2) reduction of REOs, and (3) fused salt electrolysis of rare earth chlorides or oxide–fluoride mixtures (Gupta and Krishnamurthy, 2004).

Smelting (metallothermic reduction) is the most widely used method for REM preparation. Reductants react in the furnace with oxidants (e.g., oxygen, sulfide, and carbonate) to separate and free the metal.

There are less-common processes that can be used in the liberation or reduction of rare earth compounds. These less-common processes include:

- Electrolysis
- Gaseous reduction
- Vacuum distillation
- Mercury amalgamate oxidation and reduction
- High-performance centrifugal partition chromatography
- Si-octyl phenyloxy acetic acid treatment

Potential Environmental Impacts

In 1991, the EPA identified specific waste streams in rare earth processing and assessed their hazardous waste potential (Table 11.2). The review identified four waste streams that would likely be classified hazardous: (1) waste solvent due to ignitability, (2) spent lead filter cake due to toxicity, (3) waste zinc contaminated with mercury due to toxicity, and (4) solvent extraction crud due to ignitability. However, the major environmental risk in mining and processing rare earths is associated with the treatment and disposal of the tailings (Oko-Institute e.V., 2011). The tailings typically contain high-surface-area particles, wastewater, and process chemicals. The impoundment areas are exposed to weathering conditions and have the potential to contaminate the air, soil, surface, and groundwater if not properly controlled and managed. Typical pollutants that have been associated with rare earth tailings impoundments are solids; ore associated metals (e.g., aluminum, arsenic, barium, beryllium, cadmium,

TABLE 11.2
Rare Earth Processing Waste Streams and Their Hazardous Waste Potential

Process Waste Stream	Hazardous Waste Potential
Off-gases from dehydration	None
Spent hydroxide cake	None
Spent monazite solids	None
Spent off-gases from electrolytic reduction	None (after appropriate treatment)
Spent sodium fluoride	None
Waste filtrate	None
Waste solvent	Ignitability
Spent lead filter cake	Toxicity
Lead backwash sludge	None
Waste zinc contaminated with mercury	Toxicity
Solvent extraction crud	Ignitability

Source: USEPA (1991).

copper, lead, manganese, and zinc); radionuclides; radon; fluorides; sulfates; and trace organics. Fugitive dust from the tailings impoundment can contaminate the air and surrounding soil. Surface water runoff from precipitation events or dam overtopping can transport pollutants from the impoundment to surrounding soil and surface waterbodies. Additionally, if adequate groundwater protection measures are not utilized (e.g., impoundment liner), the potential exists to contaminate surrounding groundwater resources. A worst-case scenario is dam failure due to poor construction or from a catastrophic event, resulting in serious long-term environmental damage. However, proper design, operation, and management of a mine and its associated pollution control systems can greatly reduce the risk of environmental contamination from REE mining and processing activities (USEPA, 2012).

Legacy of Environmental Damage

The production of REE has costs beyond the monetary type. For example, China's high REE production, combined with limited environmental regulations has resulted in significant environmental damage to the areas surrounding mining and processing operations. Operations range from large government-operated mines and processing facilities to small illegal endeavors. Often, smaller operations have little or no environmental controls, and larger operations have only recently begun adopting such measures. For example, after 40 years of operation, the Bayan-Oho mine has an 11-km^2 tailings impoundment that has radioactively contaminated the soil, groundwater, and vegetation of the surrounding area (Oko-Institute e.V. 2011). As reported by Hurst (1010), The Chinese Society of Rare Earths stated that every ton of rare earth produced generates approximately 8.5 kg of fluorine and 13 kg of dust. Also, they reported the use of concentrated sulfuric acid during high-temperature calcinations produces 9,600 to 12,000 m^3 of waste gas containing dust concentrate, hydrofluoric acid, and sulfur dioxide, and approximately 75 m^3 of acidic wastewater, as well as 1 ton of radioactive waste residue (Hurst, 2010). Additionally, the REE separation and refining process known as saponification had been used extensively in China until recently, generating harmful wastewater. In 2005, it was estimated that the process generated 20,000 to 25,000 tons of wastewater, with total ammonia nitrogen concentrations ranging between 300 mg/L and 5,000 mg/L (Oko-Institute e.V., 2011).

Okay, China's disregard for those things affecting the environment is one thing, how about the United States? Again, we use the Molycorp Mountain Pass site and operations as our example. Turns out that the primary source of environmental contamination at the Mountain Pass site was process wastewaters and tailings impoundments.

So, we have established that wastewaters and tailings impoundments are the areas of concern in regard to environmental contamination as a result of mining for REEs at Mountain Pass Mine. The obvious questions are what is/are mine wastewater and tailings impoundments? Good questions, for sure. Let's start with mine water. Mine water consists of all water that collects in mine workings, both surface and underground, as a result of inflow from rain or surface water and groundwater seepage. Now, depending upon the source of the water and the regional and hydrological conditions, mine water can be a significant problem because of its enormous quantity and chemical composition. Note that the mine water may have to be pumped continuously from the mine during operations. The potential impact is that mine waters can have high concentrations of heavy metals, beneficiation reagents, oils, and total dissolved solids (TDS) as well as elevated temperatures and altered pH. Because mine water is acidic, one of the most effective treatment techniques is to raise the pH by adding lime or suitable basic chemicals. Dissolved solids can be effectively removed from the water in settling ponds, adding flocculates, or "filtering" the water with ion exchange or reverse osmosis (RO). Wetlands have also been built and utilized to treat mine water. It is important to point out that the long-term effectiveness of this approach is still being studied.

With regard to mill tailings, they are the coarsely and finely ground waste potions of mined material remaining after beneficiation operations have removed the valuable constituents from the ore. Note that the physical and chemical characteristics of tailings vary according to the ore being mined and the particular beneficiation operations used. Tailings generally leave the mill as a slurry. The typical content of tailings is 50 to 70% liquid by weight and 50 to 30% solids in the form of clay, silt, and sand-sized particles. The water in the tailings impoundments may be toxic to wildlife because of the chemical reagents used in beneficiation processes and heavy metals. Leakage from tailings impoundments can also be a serious ongoing environmental problem. Leakage can transport contaminants to groundwater or surface water. Uncontrolled leakage can threaten the integrity of the impoundment structure itself, which can lead to the possibility of the catastrophic dam or embankment failure. Catastrophic impoundment failure can adversely impact downstream wildlife, aquatic organisms and their habitat, and humans.

There are occasions when tailings are de-watered prior to disposal. This is called dry tailings disposal, although the tailings may still contain water. Unlike wet tailings, which are disposed in impoundments, dry tailings are disposed primarily on large piles. These piles are non-impounding structures that make use of a variety of configurations and reduce land needs and impoundment reclamation. The major pile configurations for dry tailings are:

- *Valley-fill*—tailings are dumped to fill in a valley
- *Side hill disposal*—tailings are disposed on a side of a hill on a series of files
- *Level piles*—can grow as lift are added throughout the life of the mine

The procedure for detoxifying tailings depends on the contaminants of concern that necessitate treatment. Processes used to detoxify or stabilize cyanide and some metals include ion exchange, PH adjustment biological degradation, alkaline chlorination, and hydrogen peroxide. The gold industry usually uses chlorination and peroxide (Spellman, 2017; USEPA, 1992).

Okay, back to the United States legacy regarding REE mining and the potential environmental impacts from mining wastewater and tailings at the Mountain Pass site. Prior to 1990, the Mountain Pass site utilized onsite percolation-type surface impoundments to dispose wastewater, while convention dam impoundments were used to dispose of tailings. These past operations have impacted groundwater at the site. The most profound impact has been an increase in TDS, primarily as a result of neutralizing HCl in the wastewater with sodium hydroxide (NaOH). Groundwater TDS concentrations impacted by unlined impoundments have been reported by the current operator in the range of 10,000 mg/L. In addition, background concentrations of TDS have been documented from 360 to 800 mg/L, with low but detectable concentrations of barium, boron, strontium, and radiological constituents. Other constituents such as metals, nutrients, and radiological constituents in the wastewater and tailings have potentially had a negative impact on groundwater quality. Two additional off-site wastewater evaporation ponds were constructed in 1980 and 1987. Unfortunately, while in operation these two ponds had various failures including multiple instances of mechanical failure of the pipeline connecting the Mountain Pass site to the evaporation ponds resulted in surface soil contamination from wastewater spills. Specifically, two wastewater spills were documented at the site (USEPA, 2012). As reported by EPA, the first spill occurred in 1989 and involved the surface discharge of more than 3,000 gallons of tailings and process wastewater from a failed pipeline. In 1990, the second spill occurred involving the surface discharge of 45,000 gallons of process wastewater from a failed pipeline. The good news was that both spills were contained onsite and deemed insignificant due to the low risk to human health and the environment. The main point here is that spills can and have occurred in various mining operations and sufficient mitigation measures must be in place to ensure against harm to the public, wildlife, and the environment.

For illustration, a major point is called to be made here and that is when a spill occurs such as that at Mountain Pass Mine and Milling it is not always what you readily see on the surface that tells the whole story, so to speak. In the case of Mountain Pass wastewater spills, for instance, groundwater was contaminated—obviously, not readily seen or observed—but occur it did. So, in this case, and in the 2010 timeframe, the contaminated groundwater is actively being remediated. Groundwater interceptor wells and the mine pit have resulted in a cone of depression that is allowing capture and treatment of the contaminated plumes (CA RWQB, 2010). Additionally, the pipeline involved in the wastewater spills was removed by the former owner.

Note that subsequent strategies to minimize environmental impacts from mining operations at the Mountain Pass site include utilizing multiple technologies and management policies. The primary improvement involves the management of water and tailings. Dewatered tailings result in a "paste" that is pumped to an onsite location and layered into a stable containment mound. This process is designed to eliminate 120 acres of evaporation ponds. The plan involves employing RO to treat and reuse 90% of the wastewater, while the RO reject will be further treated to produce value-added products that can be reused in the process or sold. An onsite chlor-alkali facility will use the treated RO reject from the previous step as feed to produce NaOH, HCl, and sodium hypochlorite (NaClO) for reuse or for sale (USEPA, 2012). However, as with any reuse technology, the result is often a concentrated wastewater stream that requires disposal. In the Mountain Pass operation, pollutants such as heavy metals concentrated in the RO reject are destined for precipitation removal via nanofiltration (NF). The brine from this process is to be dried in on-site evaporation ponds prior to final disposal.

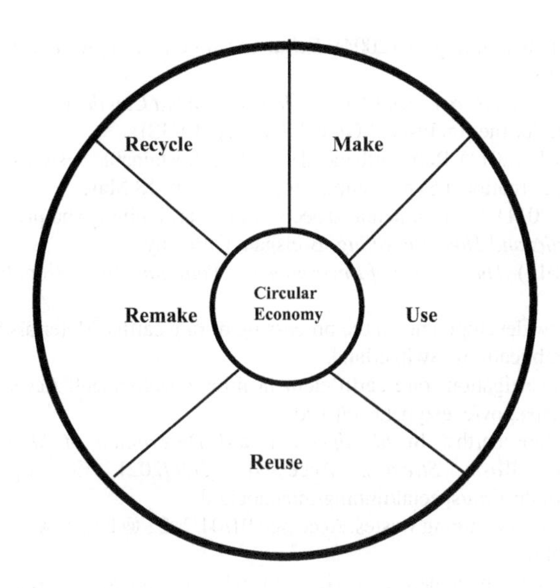

FIGURE 11.3 Circular economy approach to handling waste.

Other improvements and management strategies include (10 improved milling to improve resource recovery and lessen tailings volume per unit of REO produced, and (2) utilization of a new natural gas pipeline to supply a combined heat and power system to provide electricity and heat that is 20% more efficient than the previous methods (USEPA, 2012).

One current approach that has been put forward is called the Circular Economy approach to handling waste. In this model (see Figure 11.3), the economic system is aimed at eliminating waste and the continual use of resources. In its simplest form, the circular economy is about changing the way we produce, mine, assemble, sell and use products to minimize waste and to reduce our environmental impact. This model is also a bonus for business in that it ensures the maximizing the use of valuable resources by keeping equipment and infrastructure in use for long, thus improving the productivity of these resources. The bottom line in the circular economy is that waste materials and energy become input for other processes making the process regenerative in nature. This allows the goal to be reached in keeping the waste out of the garbage can, the holding pond, surface and groundwater systems, the oceans, and your backyard (Spellman, 2017).

REFERENCES

Aplan, F.F. (1988). The processing of rare earth minerals. Rare earths: extraction, preparation, and applications. In Proceedings of the TMS Annual Meeting, Las Vegas, Nevada, February 27–March 2, 15–34.

CARWQR (2010). China 2010 rare earth exports slip. Accessed 12/28/21 @ https://www.reuter.cor/article/usreareearth.

Ferron, C.J., Bulatovic, S.M., et al. (1991). *Beneficiation of Rare Earth Oxide Minerals.* Switzerland: Materials Science Forum. Trans. Tech Publications.

Gupta, C.K. and Krishnamurthy, N. (2004). *Extractive Metallurgy of Rare Earths*. Boca Raton, FL: CRC Press.

Hurst, C. (2010). *China's Rare Earth Elements Industry: What Can the West Learn?* Rockville, MD: Institute for the Analysis of Global Security (IAGS).

Meyer, L. and Bras, B. (2011). Rare earth metal recycling. Sustainable Systems and Technology, 2011 IEEE International Symposium, Chicago, IL, 16–18 May.

Oko-Institute e.V. (2011). Environmental aspects of rare earth mining and processing. In *Study on Rare Earths and Their Recycling*. Breisgau, Germany.

Spellman, F.R., (2017). *The Science of Environmental Pollution*, 3rd edition. Boca Raton, FL: CRC Press.

Tran, T. (1991). New developments in the processing or rare earths. Materials Science Forum, TransTech Publications, Switzerland.

USDOE (2010). Investigation: rare earth element mine development. Accessed 12/2621 @ https://nepis.epa.gov/exe/zynet.exe/p100.

USEPA (1991). Rare earths. In *Identification and Description of Mineral Processing Sectors and Waste Streams.* Accessed 12/28/2021 @ http://www.epa.gov.osw/nonhaz/industiral/special/mining/minedock/id/.

USEPA (1992). Hardrock mining wastes. Accessed 01/04/2022 @ https://www.epa.gov/npdes/pbus/wates.htm.

USEPA (2012). Rare earth elements: a review of production, processing, recycling, and associated environmental issues. EPA 600/R-12/572. Accessed 12/23/2021 @ www.epa.gov/ord.

12 Rare Earth Element Recovery and Recycling

INTRODUCTION

Expanded research and development efforts on the recovery and recycling of rare earths in waste products are currently driven by increased demand and reduced supplies along with monopolistic control practices of certain countries. At the present time (2022), commercial recycling of rare earths is very limited; however, because of increasing demand several new commercial recycling operations are in practice or are planned. The major focus on the recovery and recycling rare earths is on magnets, batteries, lighting and luminescence, and catalysts. Recycling of postconsumer, end-of-life products, typically involves the steps shown in Figure 12.1. In this chapter, a general description of each step is provided, along with the potential waste streams and environmental impacts. Note that while environmental impacts can occur, when compared with primary processing, it is reported that controlled recycling of REEs will provide significant benefits with respect to air emissions, groundwater protection, acidification, eutrophication, and climate protection (USEPA, 2012).

While research on methods for recycling and recovering REEs from as early as 1984 has been identified during the author's literature review, it is not until recently (2020–2022) that more attention within the industry and the literature has been given to the topic of REE recycling. To illustrate the current movement toward more research on recovering rare earth elements consider Case Study 12.1.

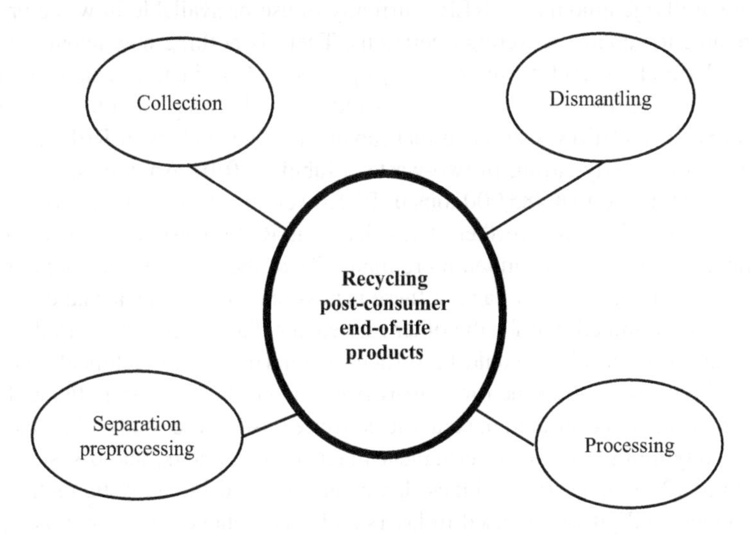

FIGURE 12.1 Steps involved in recycling of postconsumer, end-of-life products.

DOI: 10.1201/9781003350811-14

Case Study 12.1—REE Recovery from Food Waste

In 2020, EPA granted a research project (grant number SU840166) to Virginia Tech University to study rare earth elements recovery using food waste. The objective of the research aims to recover rare earths from bastnaesite ore and coal refuse using environment-friendly chemicals converted from food waste. In the performance period of this project, small molecule organic acids, such as lactic acid and succinic acid, will be converted from food waste through fermentation and/or refinement.

The acids are to be used as organic inhibitors in the concentration of bastnasite mineral from ore using from flotation. Moreover, the acids will also be tested as effective lixiviants (i.e., liquid medium used to selectively extract the desired constituent for the ore or mineral) for the recovery of rare earths from calcined and non-calcined coal refuse. Laboratory experimental results will be compiled for a detailed techno-economic assessment to investigate the merits of using organic acids, converted from food waste, as chemical reagents for REE recovery.

EPA expects the project will likely lead to an innovative and cost-effective scheme for the recovery of rare earths, which will not only contribute to food waste management but also help the United States to establish its own supply chains of rare earth elements. The human health and well-being of small, rural, tribal, and/or disadvantaged communities exposed to food waste hazards can be improved by this project in terms of reducing the amount of food waste reporting to landfills.

Again, research methods for recycling and recovering REE have been since the 1980s. The primary drivers for this research include the increased demand for REEs, concern about REE supplies, increasing cost of REEs, and new policies implemented by some countries mandating REE recycling for selected items.

There are large amounts of REEs currently in use or available in waste products that are able to support recycling operations. There is nothing new about the availability of REEs for recycling and recovery operations. Consider, for example, in 2010 a study by the Japanese government-affiliated research group National Institute for Material Science (NIMS) estimated that Japan has 300,000 tons of REEs and 6,800 tons of gold currently sitting in two-wastes (Tabuki, 2010). Another study by Yale University estimated that 485,000 tons of REEs were used globally in 2007 (Du and Graedel, 2011). The study further states that four REEs (cerium, lanthanum, neodymium, and yttrium) constituted more than 85% of the global production, and that recycling the in-use stock for each of these is possible but remains a challenge. Also, it has been determined that for the other rare earths that are generally used in much lower quantities, recycling would be difficult primarily due to technical challenges associated with separating the rare earths from the product. On an individual basis, cerium is mostly concentrated in catalytic converters (see Case Study 12.2) and metal alloys; neodymium is used in permanent magnets, computers, audio systems, cars (see Figure 12.2), and wind turbines; lanthanum is used in catalysts, metal alloys, and batteries; and yttrium is used in lasers and superconductors. To put use quantities in perspective, a Toyota Prius uses 2.2 pounds of Neodymium (Wheeland, 2010)

Case Study 12.2—REE Containing Materials and Scrap Yards

How valuable are REE-containing materials? Well, maybe we should ask Poplar Mount Baptist Church located in the tiny town of Lawrenceville, Virginia about their van. The church's van was knocked out of commission for weeks after thieves cut the catalytic converter out of its exhaust system. Several months later, across town, a catalytic convert was ripped from a van owned by First Baptist Church. The thieving of catalytic converters in town continued: a total of 15 church vans and 13 other vehicles in town, part of a nationwide surge in thefts of catalytic converters, followed.

What's going on?

It's all about easy money. Thefts of the exhaust emission control devices have increased over the past two years as prices for the precious metals they contain have skyrocketed. The going price for selling catalytic converters ranges anywhere from $50 to $300 if they are sold to scrap yards, which then sell them to recycling facilities to reclaim the metals inside, palladium, platinum, rhodium, and REEs that coat the monolith—the monolith is coated with different metals to clean emissions.

The bottom line: victims suffer the costs of replacing a stolen catalytic converter which can cost about $1,000 and make the vehicle undrivable for days or weeks as the part is ordered and installed (Lavoie, 2022).

and over 10 pounds of lanthanum (Koerth-Baker, 2010); a typical air conditioner unit includes four magnets that contain about 30 grams of rare earths (Montgomery, 2011); and a new generation windmill requires 1,500 pounds of neodymium (REVE, 2011).

E-WASTE

Note: the following information is based on the material in F. Spellman's (2021) The Science of Waste.

It doesn't actually make one of those expected and rhythmic fashion catwalks. No narrow runways or ramps are used as the flat platform running into an auditorium or between the section of an outdoor seating area, and no models demonstrate clothing during a fashion show. None of that. Okay, then are we talking about scoring an exclusive for a particular designer only. Yes, could be. Can whatever that is being featured herein or therein be considered as haute couture? Depends on your point of view with your view the emphasis. Is this item typically featured in a look book—an album of photos of the item, device, it, or whatever for fashion editors, buyers, clients, and special customers—for the viewer's perusal and to show the designer's model for the season, for the moment? How about sketches? Are there sketches of the product, the item, it, or whatever available for view, for study, for comparison?

The model, the item, it, or whatever being referred to here is high tech. The fact is tech has become another fashion item, and it can be said and easily noticed that many parts of the technology industry are now mutating into something with the anticipation and tempo of the fashion catwalk. And all this is quite confounding for

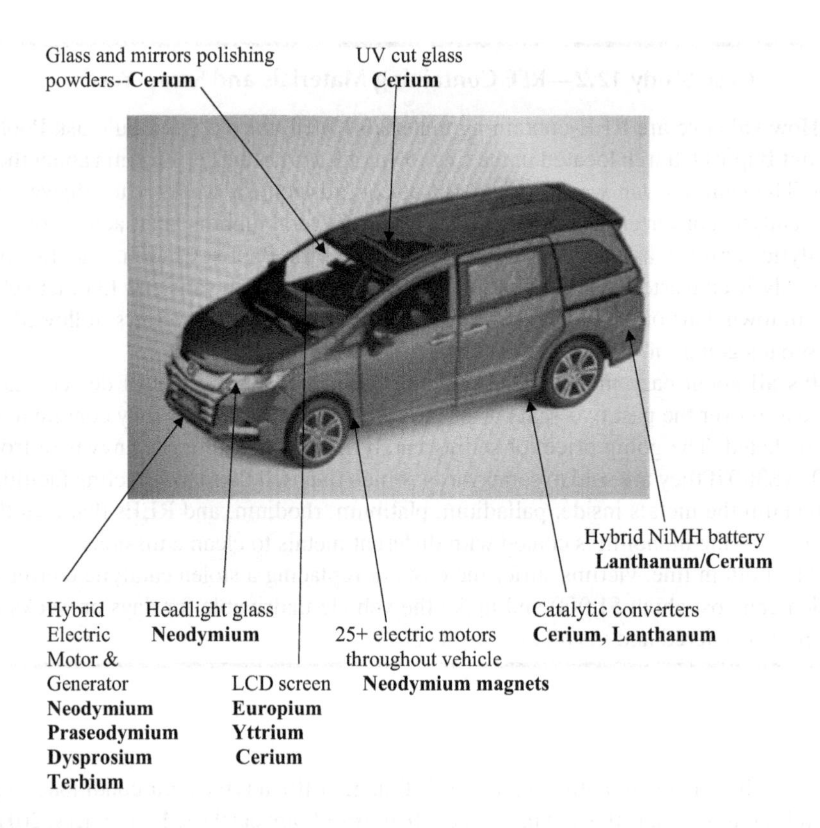

FIGURE 12.2 This figure shows a modern vehicle with some of the REEs currently used and their functions.

traditional industries to find that what is hot today is colder than an iceberg and history tomorrow. The truth be known the traditional industry personnel finds that the old rules of thumb are all of a sudden ancient and ignored.

So, what are we referring to here? What is it all about? Specifically, what is the item, device, it, or whatever? Well, be assured there are several items, devices, its, and whatever's that are being referred to here. However, clarity and to make the point transparent and as clear as a cloudless day we will focus on one item, device, it, and whatever: the mobile phone.

In the 30 years that we have enjoyed wireless mobility with our cell phones have you noticed or paid attention to the evolution of the mobile, handheld, communication devices? Probably not. Well, each new generation of mobile technology has provided more bandwidth and more possibilities. Before fast-forwarding to the present status of mobile communication devices, it is important to list a brief review or timeline of the evolution of the mobile communication devices in use at the present time. The timeline is as follows:

1983—Motorola DynaTac 8000X
1989—MicroTac

1991—Orbitel TPU 900
1997—Seimens S10 (color made available); Nokia 5100 series
1999—Nokia 7100
2000—Sharp J-SH04
2002—Sony T68i (w/camera)
2003—8100 Pearl; Sony 21020 (camera and video)
2007—Apple phone
2011–2014—Siei; Apple
2015–2018—iPhone 7 (size matters w/larger screen sizes)
Present—7 Pro devise (all about speed)
5G—ten times faster than 4G

From the list above, it is obvious that mobile communication devices have continued to evolve or upgrade to the present time. Okay, technological progress is normal, natural and to be expected. The drumbeat of innovation continues, sometimes at a startling pace. It is this startling pace of innovation that is the focus here. Simply, mobile communication devices are carry-everywhere, intimate devices. They express the user's personality, persona, and individuality. Newest, latest, slimmest, prettiest, and shiniest fastest become urgent (the key word here is "urgent") point-of-purchase carrots or inducements.

Because of the rapid advance of mobile communication devices and their associated bells and whistles, it is incumbent upon the mobile device owner to obtain and possess the latest, greatest mobile phone.

Why?

Well, maybe the purchaser of the newest mobile phone wants to be a part of the so-called in-crowd, beau monde, the beautiful people, or the neighbor next door but in this instance whether it is the nerd who invents or possesses the newest model and/or style does not matter. The latest greatest mobile phone turns the possessors of such devices into copycats because possessing the latest great device is all about fashion.

At this point, the reader may wonder what does all the preceding have to do with waste? On the other hand, maybe some of the readers understand where the preceding discussion comes into play in regard to waste. Either way, it is important to remember that we waste items when we no longer want them. We waste items when they are broken, no longer functional, or simply out of date. We also waste items when they are no longer fashionable. And that is exactly what occurs when our friends own the latest, greatest mobile phone and we do not. We simply must have the latest, greatest mobile phone—to fit in with the madding crowd and anyone else, so to speak.

The problem becomes when the old mobile phone or device is no longer in fashion. What are we to do with the old one. Unfortunately, even though still functional and expensive when purchased the dinosaur phone or electronic device must meet the same fate as the fossil. In the meantime, mobile phone users will continue to tether with their Internet connection, smartphone zombie practices, and phubbing in favor of their newest state-of-the-art mobile phones.

According to USEPA (2020) report based on data from the Consumer Electronics Association, the average American household uses about 28 electronic products such as personal computers, mobile phones, televisions, and electron readers (e-readers). With an ever-increasing supply of new electronic gadgets, Americans generated about 2.7 tons of consumer electronics goods in 2018, representing about 1% of all municipal solid waste (MSW) generation (USEPA, 2020).

LIFE CYCLE STAGES OF ELECTRONICS

Figure 12.3 shows the step-by-step life cycle of electronic.

RAW MATERIALS

Raw or virgin materials such as gold, silver, iron, palladium, copper, platinum, oil, and critical elements (REE) are found in a variety of high-tech electronics. They play crucial roles in products affecting our daily lives. These elements and materials are mined from the earth, transported, and processed. These activities use large amounts of energy and produce greenhouse gas emissions, pollution, and a drain on our natural resources. Source-reducing raw materials can save natural resources, conserve energy and reduce pollution.

The problem with e-waste is that most people have no idea with what they are wasting. If you were to ask the average person if they have any idea of what their e-device is composed of they probably will pause and scratch their heads and typically mutter something like "well, it is obvious that there is a lot of plastic involved and inside some electrical or electronic circuits and I am not sure what else is inside." It is the "I am not sure what else is inside" part of the comment that is concerning. If you were to explain that their e-device contained gold, usually $6 to $14 worth depending on the current spot price of gold, they might look at you and say, "Gee, I had no idea … but it is no matter, I have no idea how to recover the gold from the device … that is for someone else at the dump to figure out."

Sound familiar? Probably, that is if you ever ask someone or anyone who is dumping one e-device for another or whatever reason. So, what it comes down to is probably based on two factors: factor one is ignorance. The average person has no idea of the

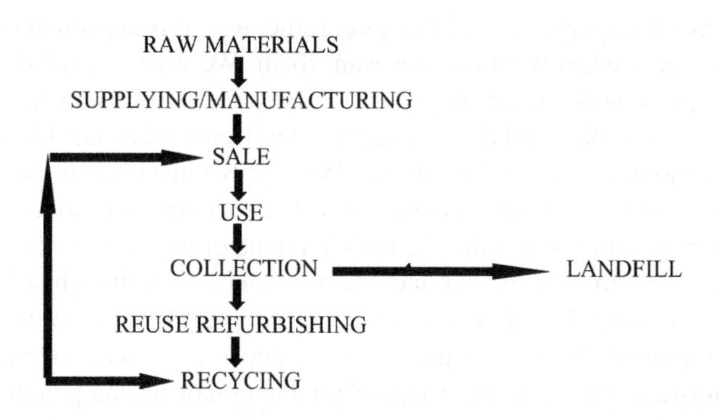

FIGURE 12.3 Life cycle stages of electronics.

Source: USEPA (2020).

value of the e-device he or she is trashing. Secondly, the person trashing the e-device could care less about the value of the e-device he or she is trashing. "My neighbor has a new one … I want one of those too." This refrain is more common than we might imagine and is one of the main reasons e-waste is a growing problem, worldwide.

The truth is few consumers of e-devices know the interior of their electronics has become thinly layered in gold; this is especially the case with computer desktops and laptops. This is why some computer makers are now offering to recycle our electronics.

You might wonder why so much of today's e-devices use gold instead of silver. Both are excellent conductors of electricity. Even though silver is less expensive than gold the problem is corrosion; simply, silver corrodes in the environment (in the open air) while gold does not. Metallic corrosion is known and predictable, especially in silver. This is the principal reason gold is used in high-end e-devices. Along with being highly corrosion -resistant gold is extremely malleable. Hence, while gold is not a better conductor of electrons than silver, it doesn't attract a sulfuric hue as silver does in the open air.

Why not use copper in e-devices; it is a lot cheaper than both gold and silver? The movement of electrons (electricity) in gold is many times faster than in silver; thus, slices of gold often get selected for high-end electronics coatings.

The bottom line, e-waste is an increasing problem. USGS (2016) estimates that only about 13% of e-waste is being recycled. The electronic waste industry has become a globalized business. About 70% of e-waste is dumped into landfills, which it is typically sorted and sold for scrap metal. Sometimes shortsightedly, this e-waste is burned to extract valuable materials; this action is harmful to people and the surrounding environment and should not be practiced.

The plain truth is that high-technology devices such as mobile phones can't exist without mineral commodities. Mined and semiprocessed materials (mineral commodities) make up more than 50% of all components in a mobile device—including its electronics, display, battery, speakers, and more (Ober et al., 2016). Figure 12.4 shows the ore minerals (sources) of some mineral commodities that are used to make components of a mobile device. Table 12.1 lists examples of mineral commodities used in mobile devices.

DID YOU KNOW?

The glass video display compound of an electronic device, a cathode ray tube (CRT), is usually found in a computer or television monitor and because it contains high enough concentrations of lead the glass is regulated as hazardous waste when disposed.

E-WASTE CAPITAL OF THE WORLD

In the Guangdong region of China, the massive electronic waste processing community of Guiyu is often referred to as the "e-waste capital of the world" (Basel Action

TABLE 12.1
Examples of Mineral Commodities Used in Mobile Devices (USGS, 2016)

Mineral Commodity	Leading Global Sources	Mineral Source(s)	Properties of Commodity
Germanium	China	Sphalerite	Conducts electricity
Graphite	China, India	Graphite	Resists heat, conducts electricity, resists corrosion
Indium	China, Korea	Sphalerite	Transparent and conducts electricity
Lithium	Australia, Chile, Argentina, China	Amblygonite, petalite, lepidolite and spodumene	Chemically reactive and has a high performance-to-weight ratio
Platinum-group metals	South Africa, Russia, Canada	More than 100 minerals	Conducts electricity
Potassium	Canada, Russia, Belarus	Langbeinite, sylvite, sylvinite	Strengthens glass
Rare-earth elements	China	Bastnaesite, ion adsorption clays, loparite, monazite, and xenotime	Highly magnetic; blue, green, red, and yellow phosphors, and optical-quality glass
Industrial sand	China, United States	Silica sand	Give glass clarity
Silicon	China	Quartz	Conducts electricity
Silver	Mexico, China, and Peru	Argentite and tetrahedrite	Circuitry and conducts electricity
Tantalum	Rwanda, Brazil, Congo	Columbite and tantalite	Stores electrical charge
Tin	China, Indonesia, Burma, Peru	Cassiterite	Transparent and conducts electricity
Tungsten	China	Scheelite and	Highly dense and durable for vibrator's weight component

Network, 2002; Roebuck, 2012; Slade, 2006). In the past, Guiyu was an agriculture community; however, in the mid-1990s it transformed into an e-waste recycling center involving over 75% of the local households and an additional 100,000 migrant workers (Wong, 2007). Laborers are employed in thousands of individual workshops to snip cables, pry chips from circuit boards, grind plastic computer cases into particles, and dip circuit boards in acid baths to dissolve the precious metals. Others work to strip insulation from all wiring in an attempt to salvage tiny amounts or copper wire (EWDW, 2012). Practices within these workshops such as uncontrolled burning, disassembly, and disposal have led to a number of environmental problems such as groundwater contamination, atmospheric pollution, and water pollution either by immediate discharge or from surface runoff (especially near coastal areas), as well as health problems, including occupational safety and health effects among those directly and indirectly involved, due to the methods of processing the waste.

Situated on the South China Sea coast, Guiyu consists of several villages. Six of these villages specialize in circuit-board disassembly, seven in plastics and metals

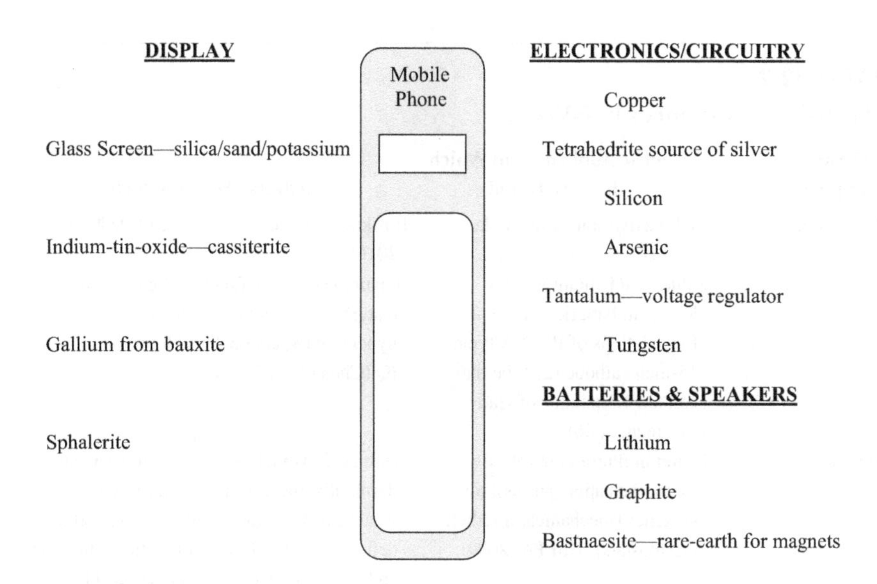

FIGURE 12.4 Minerals and derivatives in mobile phones.

Source: USGS (2016).

reprocessing, and two in wire and cable disassembly. The environmental group Greenpeace sampled dust, soil, river sediment, and groundwater in Guiyu. They found very high levels of toxic heavy metals and organic contaminants (Seattle Times, 2012). The burning off of plastics in the region has resulted in 80% of its children having dangerous levels of lead in the blood (Monbiot, 2009). One campaigner for the group found over 10 poisonous metals such as lead, mercury, and cadmium.

Sadly, Guiyu is only one example of digital dumps and similar places that are currently found across the world in Nigeria, Ghana, and India (Greenpeace, 2015). Table 12.2 lists e-waste components and where they are found in electric/electronic appliances or devices and their adverse health effects.

RECYCLING SITE NEIGHBORS

Residents living around the e-waste recycling sites, even those who are not involved with the recycling, can face the environmental exposure due to the food, water, and environmental contamination caused by e-waste because they can easily be exposed to or in contact with the e-waste contaminated air, water, soil, dust, and food sources. In general, there are three main exposure pathways: inhalation, ingestion, and dermal contact (Grant et al., 2013).

Studies show that people living near e-waste recycling sites have a higher daily intake of heavy metals and a more serious body burden (toxic load). Potential health risks include mental health, impaired cognitive function, and general physical health damage (Song and Li, 2015). DNA damage was also found more widespread in all the e-waste exposed populations (i.e., adults, children, and neonates) than the

TABLE 12.2
Hazardous Substances in E-Waste

E-Waste Component	Electric Appliances in Which They Are Found	Adverse Health Effects
Americium	Radioactive source in smoke alarms	It is known to be carcinogenic (TOXNET, 2016)
Lead	Solder, CRT monitor glass, lead-acid batteries, some formulations of PVC. A typical 15-inch cathode ray tube may contain 1.5 pounds of lead (Morgan, 2006).	Impaired cognitive function, behavioral disturbances, attention deficits, hyperactivity, conduct problems, and lower IQ (Chen et al., 2011).
Mercury	Found in fluorescent tubes (numerous applications), tilt switches (mechanical doorbells, thermostats) (USEPA, 2009).	Health effects include sensory impairment, dermatitis, memory loss, and muscle weakness. Exposure in-utero causes fetal deficits in motor function function, attention, and verbal domains (Chen et al., 2011).
Cadmium	Found in light sensitive resistors, corrosion-resistant alloys for marine and aviation environments, and nickel-cadmium batteries.	The inhalation of cadmium can cause severe damage to the lungs and is known to cause kidney damage (Lenntech, 2014).
Hexavalent chromium	Used in metal coatings to protect from corrosion.	A known carcinogen after occupational inhalation exposure (Chen et al., 2011).
Sulfur	Found in lead-acid batteries.	Health effects include liver damage, kidney damage, heart damage, eye and throat irritation. When released into the environment, it can create sulfuric acid through sulfur dioxide.
Brominated flame retardants	Used as flame retardants in plastics in most electronics.	Health effects include impaired development of the nervous system, thyroid problems, and liver problems (Birnbaum and Staskal, 2004).
Perfluorooctanoic acid	Used as an antistatic additive in Industrial applications and found in Non-stick cookware.	Studies in mice have found the following health effects: Hepatoxicity, developmental toxicity, immunotoxicity, hormonal effects, and carcinogenic effects (Wu et al., 2012).
Beryllium oxide	Filler in some thermal interface materials such as thermal grease used on heatsinks for CPSs and power transistors (Becker et al., 2005) magnetrons, X-ray-transparent ceramic windows, heat transfer fins in vacuum tubes, and gas lasers.	Occupational exposures associated with lung cancer, other common adverse health effects are beryllium sensitization, chronic beryllium disease, and acute beryllium Disease (OSHA, 2016).

(continued)

TABLE 12.2 ***Continued***
Hazardous Substances in E-Waste

E-Waste Component	Electric Appliances in Which They Are Found	Adverse Health Effects
Polyvinyl chloride (PVC)	Commonly found in electronics and is typically used as insulation for electrical cables (Greenpeace, 2021)	In the manufacturing process, toxic and hazardous raw materials, including dioxins are released. PVC such as chlorine tend to bioaccumulate (Electronicstakeback, 2021).

populations in the control area (Song and Li, 2015). Experience has shown that DNA breaks can increase the likelihood of wrong replication and thus mutation, as well as lead to cancer if the damage is to a tumor suppressor gene (Liulin et al., 2011).

RECYCLING REES

In 2011, the United Nations reported on the recycling rates of metals estimate that the end-of-life functional recycling (i.e., recycling in which the physical and chemical properties that made the material desirable in the first place are retained for subsequent use) for rare earths was less than 1% (UNEP, 2011). Another 2011 study estimated that world-wide, only 10% to 15% of personal electronics are being properly recycled (Dillow, 2011). Of the items that are sent for recycling, the European Union (EU) estimates that 50% of the total is illegally exported, potentially ending up in unregulated recycling operations in Africa or Asia. These recycling operations frequently result in environmental damage and worker exposure, as documented in a separate UNEP report (Schluep et al., 2009).

Urban Mining

The increasing world demand for REEs and the accompanying increase in prices (along with other recyclable metals) and with the knowledge of the quantities available in discards have led to the concept of "urban mining" (aka "urban prospecting"). Simply put, the main driver of urban mining is simple enough: increased demand and reduced supply of REEs, along with the quantities available in waste products. Consumer electronics are increasingly becoming subject to urban mining practices—and with over 7 million metric tons (and counting) of personal computers, computer monitors and peripherals, televisions, and mobile devices being generated in the United States, Europe, China, and India—the supply available for "mining," or recycling, is large (very) and increasing in digital leaps and bounds, so to speak.

Typically, small electronics such as cell phones reach their end of life after a few years, while many of the products that contain larger amounts of REEs have useful lives of well over one decade, or so. The problem is that the recycling of REEs contained within products will occur many years in the future and is not a short-term solution to the current demand. Another problem or consideration that will probably impact one of the recycling drivers is that as additional REE mines begin operation outside of China, global production will increase, costs may decrease, and the urgency behind the push to recycle may be reduced.

RECYCLE PROCESSING STEPS

As previously note and illustrated in Figure 12.1, the recycling process for post-consumer, end-of-life products typically involves four key steps: collection, dismantling, separation (preprocessing), and processing.

In a dated UNEP (2011) status report, it stated that the end-of-life recycling rate, defined as the "percentage of a metal in discards that is actually recycled," for REEs is less than 1%. As cited by Meyer and Bras (2010), the consumer products with the rarest earth recycling potential are the ones that contain high levels of rare earths and an established collection or recycling infrastructure, such as fluorescent lamps, magnets, car batteries, and catalytic converters. UNEP (2011) noted that three factors contributing to the effectiveness of recycling efforts are the following:

1. Economics—the value of the materials to be recycled must be greater than the cost of recycling. In situations where this is not the case, laws and incentives can be effective in increasing recycling rates.
2. Technology—products that are designed with recycling in mind will be easier to dissemble and re-process.
3. Societal—programs will be more effective when the public is aware of the benefits of recycling and the collection and recycling infrastructure is accessible and well publicized.

Recycling can be conducted on either pre-consumer or post-consumer items. While experience has that most e-waste material recycling pertains to post-consumer actions, it has been reported that 20 to 30% of rare earth magnets are scrapped during the manufacturing process and ways for recycling this "waste stream" (Schuler et al., 2011; Spellman, 2017).

Collection

The first step, defined as collection, is discussed in a report by UNEP (Schluep et al., 2009) and more recently by Spellman (2022) and can be accomplished through a variety of means, but it is generally more efficient when a collection infrastructure is already established. In the past, state regulations have been effective at establishing the collection and recycling infrastructure required to increase recycling rates for consumer goods in the United States. For example, in the 1980s, the recycling of lead-acid car batteries became required by many states and resulted in a 95% recycling rate by 1990. In 1990, EPA data show that nationally 19% of consumer electronics were recycled in 2009 (Bomgardner, 2011). Presently, several states have laws requiring e-waste recycling.

The EPA's Plug-In To e-Cycling Partners website provides links to tack-back-programs and drop-off locations for mobile devices, computers, printers, and televisions. The partners include retail stores, equipment manufacturers, and mobile device service providers. Together, they collected and recycled several million pounds of used consumer electronics. Collection methods include direct mail of products to locations established by equipment manufacturers and drop off of used products at designated locations, such as retail stores or locations specifically setup as part of collection day events. Environmental impacts from the collection step are most predominantly due to transportation/shipping of materials to the collection point and from the point of collection to the location of the processing facility (Spellman, 2022; USEPA, 2012).

Dismantling and Preprocessing

Dismantling and preprocessing steps are vital for separating high-value components from less valuable materials. Often, high-value materials such as REEs and other metals, like gold and silver, make up a small percentage of the item being recycled, and separation steps will make it more efficient to recover them. Nevertheless, even when metals are separated from other nonmetal components, mixed metal scrap is more challenging to recycle segregated metals.

Standard dismantling and preprocessing steps include manual or mechanical separations, manual or mechanical disassembly, mechanical shredding, and screening. Note that advances to manual separation methods are on-going and continuously investigated as a way to reduce costs, increase speed, reduce damage to selected materials desired for removal, and reduce potential worker exposures. Two early examples of successes in this area are the following:

- Hitachi developed a process for dissembling hard disk drives that involve placing the drives in a rotating drum where forces such as shock and vibration are employed. This process is reported to be eight times faster than manual separation and, therefore, more cost-effective (USEPA, 2012).
- USEPA (2012) reported that a process developed by NIMS that includes a small-scale electronic crushing device that, in a few seconds, is able to reduce cell phones and small home appliances to small pieces. This step is followed by placing the pieces in a three-dimensional ball mill that degrades the parts recovered from the crushing device to a powdered form. Because of the short treatment time in the ball mill, the remaining pieces of plastics and other materials remain intact and therefore can be recovered in a condition that allows for plastic recycling. The powder can then be further processed to recover metals of concern (USEPA, 2012).

It is important to point out that during the dismantling and preprocessing steps, hazardous or other unwanted substances have to be removed and then either stored or treated safely while valuable materials are removed for reuse or recycling. For devices containing ozone-depleting substances, such as refrigerators and air-conditioners, the degassing step is crucial in the preprocessing stage because the refrigerants used (e.g., chlorofluorocarbon or hydrochlorofluorocarbon), in older models, need to be removed carefully to avoid air emissions of ozone-depleting substances, which have a large global climate potential. LCD monitors containing mercury or other toxic metals need to be dismantled with care to ensure with care to ensure worker and environmental protection. Circuit boards present in electronic equipment can contain lead in solder and flame retardants containing resins (Schluep et al., 2009). After removal of hazardous and other special components, the remainder of the item being recycled can be further separated by manual dismantling or mechanical shredding and (automated) sorting techniques. Some shredding technologies have the potential to generate dust or other particulate matter that can impact worker health. Additionally, all mechanical processing equipment requires energy inputs that have additional associated environmental impacts.

Processing

After completion of the preprocessing steps, the components of interest are ready for the processing step. Processing techniques that are currently used or are in the development or research stage for the recovery of REEs are listed below, but keep in mind that the development of other processes is a work in progress (USEPA, 2012).

- Pyrometallurgy processes are energy intensive, using high temperatures to chemically convert feed materials and separate them so that the valuable metals can be recovered.
- Hydrometallurgy processes use strong acidic or basic solutions to selectively dissolve and then precipitate metals of interest from a preprocessed powder form.
- Electrometallurgy processes such as electrowinning (where a current is pass from an inert anode through a liquid stripping solution containing the metal; the metal is extracted by an electroplating process, which deposits the rate earths onto the cathode), and electrorefining (where the anode is composed of recycled material—when the current passes from the anode to the cathode through the acute electrolyte, the anode corrodes, releasing the rare earth ions into the solution, then electrowinning occurs).
- Dry processes stage uses hydrogen gas at atmospheric pressure to turn neodymium-containing magnets to a powder that can then be re-formed into new magnets under heat and pressure (Davies, 2011).
- Tailings recycling involves reprocessing of exiting tailings to recover the remaining amounts of REEs they contain.
- Microbe-filled capsule technology where capsules are placed in a medium containing rare metals, and the microbes then absorb the metals.
- Titanium dioxide process where recovery rates of oxides of neodymium, cerium, and lanthanum vary between 50% and 80% and are thought to be able to be increased in the future (University of Leeds, 2009).

COMMERCIAL REE-RECYCLING

The number of commercial REE-recycling operations is limited but that is rapidly changing due to demand for REE and other critical metals/minerals. To date, the focus of REE commercialization has been on magnets, batteries, lighting and luminescence, and catalysts (Schuler et al., 2011; Spellman, 2022). It should be noted that in many cases the companies that are in the process of using or developing REE-recycling technologies have not published reports or papers with the details of their individual processes, as there is generally a competitive advantage, considered proprietary, or trade secrets.

ENVIRONMENTAL IMPLICATIONS OF RECYCLING REES

Uncontrolled recycling of e-wastes has the potential to generate significant hazardous emissions (Schluep et al., 2009; Spellman, 2017, 2022). While this section is focused on REE, and e-wastes, the emission categories presented herein pertain to the recycling of other types of wastes as well.

1. **Primary emissions**—hazardous substances contained in e-waste (e.g., lead, mercury, arsenic, polychlorinated biphenyls (e.g., lead, mercury, arsenic, polychlorinated biphenyls [PCBs], and ozone-depleting substances).
2. **Secondary emissions**—hazardous reaction products that result from improper treatment (e.g., dioxins or furans formed by incineration/inappropriate smelting of plastics with halogenated flame retardants).
3. **Tertiary emissions**—hazardous substances or reagents that are used during recycling (e.g., cyanide or other leaching agents) and are released because of inappropriate handling and treatment. UNEP reports this is the biggest challenge in developing countries engaged in small-scale and uncontrolled recycling operations (Schluep et al., 2009)

It is important to point out that for recycling operations using pyrometallurgy, facilities need to have regulated gas treatment technologies installed and properly operating to control VOCs, dioxins, and other emissions that can form during processing. Also, for hydrometallurgical plants, special treatment requirements are necessary for the liquid and solid effluent streams to ensure environmentally sound operations and to prevent tertiary emissions of hazardous substances.

Recovering metals from state-of-the-art recycling processes is reported as being two to ten times more energy efficient than smelting metals from ores. Recycling also generates only a fraction of the CO_2 emissions and has significant benefits compared to mining in terms of land use and hazardous emissions. While examples are not provided in the reports specifically for REEs, production of 1 kg aluminum by recycling uses only one-tenth or less of the energy required for primary production and prevents the creation of 13 kg of bauxite residue, 2 kg of CO_2 emissions, and 0/011 kg of SO_2 emissions. For precious metals, the specific emissions saved by state-of-the-art recycling are reported as being even higher (Schluep et al., 2009).

Various reports including one by Schuler et al. (2011) stated that when compared with primary processing, recycling of REEs will provide significant benefits with respect to air emissions, groundwater protection, acidification, eutrophication, and climate protection. An additional benefit of recycling REEs is that it will not involve radioactive impurities, as is the case with primary production.

There are additional benefits of recycling that are not directly linked with the environment and the main one is an improved supply of REEs, and therefore less dependence on foreign sources; the potential for reduction in REE costs due to supply increases and reduction of the current "monopoly" from foreign suppliers; and the potential for job creation from an expanded recycling industry.

THE BOTTOM LINE

Rare earth elements or minerals are not really all that rare, but they are critical to modern life and society. What is rare, however, is the science and technology. That is, the science and technology needed not only to allow the mining of REE effectively and efficiently but also the science and technology needed for recovery and recycling of REE and other critical metals is rare. Very limited recycling of these critical elements currently takes place. Advances can be made in the recycling of the REE from magnets, fluorescent lamps, batteries, catalysts, and other devices. Increased amounts of REE recycling are needed to ensure security. Also, recycling of the REE can also balance the problems related to primary REE production. Lastly, and most importantly, science and technology must step in to ensure the protection of the environment, humans involved in recovery and recycling operations, and the ecosystem in general. In the next chapter, I make a point of addressing not only environmental and human concerns related to REE production, recovery, and recycling of REEs.

REFERENCES

Basel Action Network (2002). Exporting harm: the high-tech trashing of Asia. Accessed 02/19/2021 @ http://ban.org/E-waste/technotrashfinalcomp.pdf.

Becker, G., Lee, C. and Lin, Z. (2005). Thermal conductivity in advanced CHIPS: emerging generation of thermal greases offers advantages. Accessed 02/11/2021 @ https://web.archive.org/web/20000621233638/http:www.apmag.com/.

Birnbaum, L.S. and Staskal, D.F. (2004). Brominated flame retardants: cause for concern? *Environmental Health Perspectives* 112(1): 9–17.

Bomgardner, M. (2011). Taking it back. Material makers will have to adapt to help consumer goods firms fulfill product stewardship goals. *Chemical and Engineering News* 89: 31, 3–17.

Chen, A., Dietrich, K.N. and Huo, X. (2011). Developmental neurotoxicants in E-waste: an emerging health concern. Accessed 02/11/2021 @ https://ncbi.nim.nih.gov/pmc/acticles/PMC3080922.

Davies, E. (2011). Critical thinking. *Chemistry World*, January. Accessed 01/10/2022 @ www.chemistryworld.com.

Dillow, C. (2011). A new international project aims to back U.S. electron waste for recycling. Accessed 01/10/2022 @ www.popsci.com.

Du, X. and Graedel, T. (2011). Global in-use stocks of the rare earth elements: a first estimate. American Chemical Society. *Environmental Science & Technology* 45: 4096–4101.

Electronicstakeback (2021). Flame retardants & PVCs in electronics. Accessed 02/11/2021 @ http://www.electronicstakeback.com/toxics-in-electroncis/flame-retardants-pvc-and-electroncs.

EWDW (2012). Electronic waste dump of the world. Accessed 02/19/2021 @ http://sometimes-interesting.com/2011/07/17/electronic-waste-dump-of-the-world/.

Grant, K., Goldizen, F.C., Sly, P.D., Brune, M.-N., Neira, M., van den Berg, M. and Norma, R.E. (2013). Health consequence of exposure to E-waste: a systematic view. *The Lancet Global Health* 1(6): 350–361.

Greenpeace (2021). Why BFRs and PVC should be phased out of electron devices. http://www.greenpeace.org/archive-international/en/campaigns/detox/electronics/the-e-waste-problem/what-s-in-electronc-devices/bfr-pvc-toxic/.

Koerth-Baker, M. (2010). 4 rare earth elements that will only get more important. *Popular Mechanics*, May 21. Accessed 01/10/2022 @ http://www.Popularmechanics.com/technology/engineering/news/important-rare-earth-elements.

Lavoie, D. (2022). *Catalytic Converters Targeted. In the Virginia-Pilot.* Norfolk, VA: Associated Press.

Lenntech (2014). Cadmium (Cd)-chemical properties, health and environmental effects. Accessed 01/22/2021 @ http://web.archive.org/web/20140515101400/http://www.lenntech.com/periodic/elements/cd.htm#ixzz1MpuZHWfr.

Luilin, S. et al. (2011). Tumor suppression genes. Accessed 12/15/21 @ https://pubmed.ncib.nim.nih.gov/1756265.

Meyer, L. and Bras, B. (2011). Rare earth metal recycling. Sustainable Systems and Technology, 2011 IEEE International Symposium, 16–18 May, Chicago, IL.

Monbiot, G. (2009). From toxic waste to toxic assets, the same people always get dumped on. *The Guardian*. London. Accessed 02/18/2021 @ https://www.theguardian.com/commentisfee/cif-green/2009/sep/21/global-fly-tipping-toxic-waste.

Morgan, R. (2006). Tips and tricks for recycling old computers. Accessed 02/19/2021 @ http://www.smartbiz.com/acrticle/articleprint/1525-1/58.

Montgomery, M. (2011). Rare earth recycling, enter a new reality. *IB times.com*, February 23. Accessed 01/10/2022 @ http://www.Ibtimes.com/articles/115379/20110223/rare-earth-recycling-enter-the-new-reality.htm.

Ober, S. (2016). Global metal ore production. Accessed 12/21/21 @ https://www.researchgate.net/figure/globalmetal.

OSHA (2016). Health effects. Accessed 02/02/2021 @ https://www. osha.gov/SLTC/beryllium/healtheffects.html.

REVE (Regulación Eólica con Vehívulos Eléctricus) (2011). Rare earths and wind turbines. May 16. Accessed 01/10/2022 @ htt://www.evwind.es/notice.php?id_not=11586.

Roebuck, K (2012). Electronic waste: high-impact strategies. Accessed 02/19/2021 @ https://books.google.com/books?id=RjYPBwAAQBAJ&q=Activists+push+for+safer+e-recyclin&pg=PA8.

Schluep, M., Hagelueken, C., et al. (2009). Recycling from E-waste to resources. United Nations Environment Programme & United Nations University, July. Accessed 01/11/2022 @ http://www.unep.org/PDF/PressReleases/E-Waste_publication_screen_FINALVERSION-sml.pdf.

Schuler, D., Buchert, M., et al. (2011). Study on rare earths and their recycling. OKO-insititut e v., January.

Seattle Times (2012). E-waste dump of the world. Accessed 02/19/2021 @ http://seatteltimes.com/htiml /nationworld/2002920133_ewaste09.html.

Slade, G. (2006). Computer age leftovers. Accessed 02/19/2021 @ http://www.denverpost.com/perspective/ci_3633138.

Song, Q. and Li, J. (2015). A review of human health consequences of metal exposure to E-waste in China. *Environmental Pollution* 196: 450–461.

Spellman, F.R. (2017). *The Science of Environmental Pollution*, 3rd ed. Boca Raton, FL: CRC Press.

Spellman, F.R. (2022). *The Science of Waste.* Boca Raton, FL: CRC Press.

Tabuki, H. (2010). Japan recycles minerals from used electronics. *The New York Ties*, October 4. Accessed 01/11/2022 @ http://www.nytimes.com/2010/10/05/business/global/05recycle.html.

TOXNET (2016). Americium, radioactive. Accessed 02018/2021 @ http://toxnet.nim.nih.gov/-cgi-bin/sis/search/a?dbs+hsdb:@DOCNO+7383.

UNEP (2011). Annual Report/UNEP. Accessed 12/15/21 @ https://www.unep.org/resource annualreport/org.

University of Leeds (2009). Rare earth recovery. Accessed 12/12/21 @ https://www.theengineer.co.uk/content/news/rare-earth-oxide-recovery.

USEPA (2009). Question 8. Accessed 02/05/2021 @ http://www.epa.gov/dfe/pubs/comp-dic-ica-sum/ques8.pdf.

USEPA (2012). Rare earth elements: A review of production and processing. Accessed 12/21/21 @ https://nepis.epa.gov/exe/zyzynet.cgi.

USEPA (2020). *Rare Earth Elements Recovery Using Food Waste*. EPA Grant Number: SU840166. Washington, DC: United States Environmental Protection Agency.

USGS (2016). *Obsolete Computers, "Gold Mine," Or High-Tech Trash?* Washington, DC: U.S. Department of Interior—U.S. Geological Survey.

Wheeland, M. (2010). Rare earth prices spike, relief may be yours away. *GREEN BIZ.COM*, May 5. Accessed 01/11/2022 @ http://www.greenbiz.com/news/2011/05/05/rare-earth-prices-spike-relief-may-be-years-away.

Wong, M.H. (2007). Export of toxic chemicals–a review of the case of uncontrolled electronic-waste recycling. Accessed 02/19/2021 @ https://repository.hkbu.edu.hk/cgi/viewcontent.cgi?article=1000&context=cies_ia.

Wu, K., Xu, X., Peng, L., Liu, J., Guo, Y. and Huo, X. (2012). Association between maternal exposure to perfluorooctanoic acid (PFOA) from electron waste recycling and neonatal health outcomes. *Environmental International* 41: 1–8.

13 REE
Human Health and Environmental Risks

To this point in the book the potential benefits of REEs—many of which we are not even aware of yet but with advances in research and technology it is certain that additional benefits will be derived from using REEs—have been discussed. Well, it must be said that there is usually two sides to each coin and two edges to many swords and with REEs there are two sides or edges: a beneficial side and a risky side. Environmental and health issues occupy the risky side of the coin or the other risky, sharp edge of the sword. Exposure to REEs during production, processing, and recycling can potentially lead to adverse health effects in humans, animals, and plants, including REE bioaccumulation and REE-induced pathologies. In this chapter we present the potential risks of production, processing, and recycling of REEs.

—*F. Spellman (2013)* The Handbook of Environmental Health

INTRODUCTION

Since the 1990s, EPA has performed a number of studies (many are on-going) to evaluate the environmental risks to human health and the environment from hardrock mining and metal ore processing activities. As with many other operations, such as fracking, the most significant environmental impact from contaminant sources associated with hard rock mining is to surface water and groundwater quality. However, impacts have also been reported to sediments, soils, and air. Note that relative comparisons can be made for mining of rare earth mineral ores and processing those ores into the final products with other hard rock metal mining and processing operations. While every deposit is geochemically unique and every mine processor must conform to the characteristics of the ore deposit it is still possible and plausible to make these comparisons. The environmental and human health impacts are largely associated with the release of mine waters that typically contain elevated concentrations of metals, industrial chemicals used to maintain the mine site and equipment, and processing chemicals needed for milling and final processing steps. Keep in mind that at the present time the specific health of elevated concentrations of REEs in the environment, like forever chemicals found in today's treated wastewater effluents, are not well understood. Moreover, there is also potential for impacts to human health and the environmental from recycling activities to recover REMs. The truth be known, we simply do not know what we do not know about the environmental and human

DOI: 10.1201/9781003350811-15

health impacts of mining, processing, and recycling REEs. This chapter presents a general conceptual site model (CSM) for a generic, aboveground hard rock mine site.

Again, mining and processing activities have the potential to create a number of environmental risks to human health and the environment. Note that the severity of these risks is highly variable between mine and mine plant operations. The contaminants of concern will vary depending on the REE-mineral ore, the toxicity of the contaminants from the waste rock, ore stockpiles, and process waste streams. With regard to contaminants, it is all about their mobility. The mobility of contaminants will be controlled by the characteristics of the geologic, hydraulic, and hydrogeologic environments where the mine is located, along with the characteristics of the mining process and waste handling methods. It is important to point out that the environmental impact from urban mining of REM recycling operations is similar to mineral processing, since recovery and refining methodologies can be identical (USEPA, 2012).

A summary of potential emission points and pollutants of concern associated with mining, processing, and recycling REEs is presented in Table 13.1.

CONCEPTUAL SITE MODEL

This section provides a generic CSM to illustrate and provide a general perspective for common sources of contamination along with typical contaminant release, transport, and fate scenarios that could be associated with a larger hard rock mine site and operation. The model describes the source(s) of contaminants, the mechanisms of their release, the pathways for contaminant transport, and the potential for human and ecological exposure to chemicals in the environment. In plainer English, what the CSM provided here does is to show you the source, fate and transport, and exposure pathways and receptors/resources we're interested in protecting. Keep in mind that the CSM is dynamic and its development is iterative (i.e., repetitive)—based on on-going investigation findings. Moreover, the CSM helps us prioritize investigation as it advances and help us target media or specific pathways that appear to drive risk or hazard. Also, the CSM, again, is not static it will grow and shrink as it moves through these uses. And the CSM can be reflective of broader areas, with "grayed Out" areas or pathways which are not priority issues.

Anyway, as in other parts of this book, the discussion presented herein is general and can't specifically address every circumstance or condition (gee, can any work do that?). However, to reiterate a precious point: while the geochemistry of the ore, and therefore the characteristic of pollutants, is likely quite different, a REE mine and operation is similar to many other hard rock mining operations and the methods used to beneficiate and ill REE-mineral ores are also similar. While the basic metallurgical processes used to extract a metal from hard rock mineral ores are similar, the actual mineral ore processing steps used to recover a metal or metal oxide are varied. Metallurgical processing is generally unique to the deposit's geochemistry, and therefore the actual methods and chemicals used are often proprietary. Environmental impact occurs at every stage of the mine's life-cycle (USEPA, 2012).

Okay, the CSM presented in this discussion assumes that within the property boundaries of a single mine it is important to note that there can be a variety of

TABLE 13.1
Summary Table of Pollutants, Impacted Environmental Media, Emission Sources, and Activity Associated with REE Mining, Processing, and Recycling

Activity	Emission Source (s)	Primary Pollutants of Concern
Mining (aboveground and underground methods)	Overburden	Radiologicals
	Waste rock	Metals
	Sub-ore stockpile	Mine influenced waters/acid mine
	• Ore stockpile	Drainage/alkaline or neutral mine
		Drainage
		Dust and associated pollutants
Processing	Grinding/crushing	Dust
	Tailings	Radiologicals
	Tailings impoundments	Metals
	Liquid waste from processing	Organics
		Turbidity
		Dust and associated pollutants
Recycling	Collection	Dust and associated pollutants
	Dismantling and Separation	VOCs
	Scrap waste	Metals
	Landfill	Organics
	Processing	Dust and associated pollutants
		VOCs
		Dioxins
		Metals
		Organics

Source: USEPA (2012).
Note: Again, limited toxicological or epidemiological data are available to assess the potential human health effects of REEs. Studies are on-going to determine the human and ecological effects of mining and processing REEs. What is reported in this work is what we know now.

support process areas and facilities. We are talking about what usually is a complex operations with various unit processes working in harmony to mine and process ore. The truth be told there are waste materials that are associated with each step of mining and the subsequent ore processing steps used in processing metallurgy for the target metal. Most of the time, the waste streams from mining and mineral processor can include sediment, particulates, vapors, gases, wastewater, various chemical solvents, and sludge from chemical extraction and filtration steps. Many mining and processing operations will produce these and/or other wastes that required management and have the potential to create environmental risks to human health and sensitive habitat—and this is what this chapter and our discussion are all about.

Note that the CSM exposure pathways provided in Figure 13.1 represents the goings-on at a typical mining and processing site—a typical one—of nonspecific location, climate, and physical setting. There are several factors that influence the likelihood and potential severity of associated risks associated with a specific mine

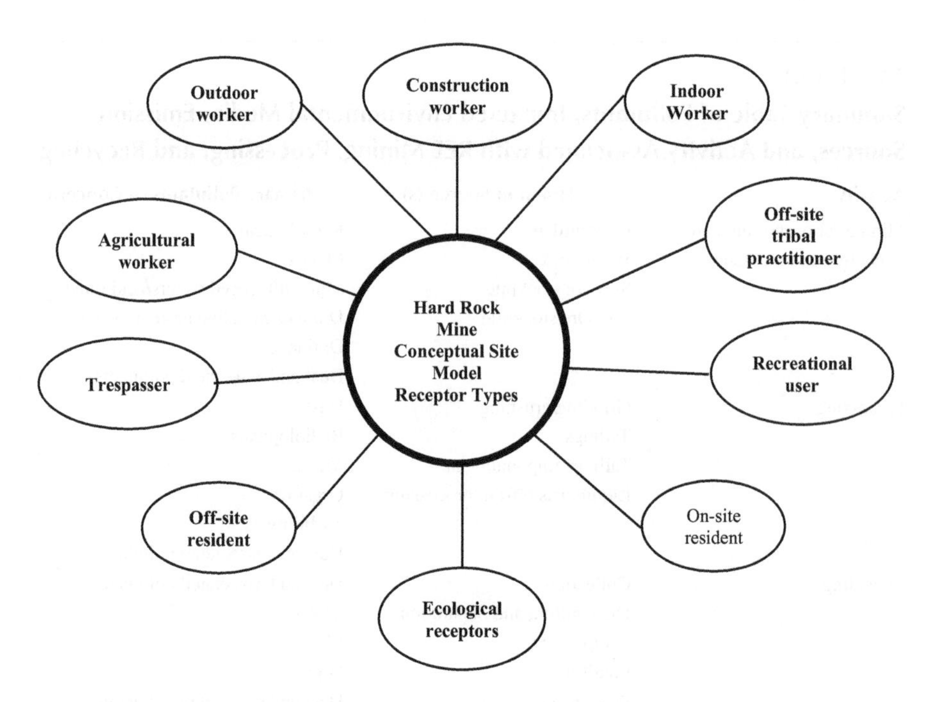

FIGURE 13.1 Hard rock mine conceptual site model exposure pathways and chemical receptors (USEPA, 2009).

site include mine conditions, ore geochemistry, and geologic, physiographic, and hydrogeological settings. I have provided this CSM for the reader to gain a general perspective and to orient the reader who is unfamiliar with the site features and conditions typically found at a commercial hard rock mine site. The features included in the CSM include mine pits, leach piles, other processing areas, tailings, and waste piles.

Okay, let's get to it. As previously discussed in this book, mining is the removal of ore from the ground on a large scale by one or more or four principal methods: surface mining, underground mining, place mining, and in situ mining (extraction of ore from a deposit using chemical solutions). After the ore is removed from the ground, it is crushed so that the valuable mineral in the ore can be separated from the waste material and concentrated by flotation (a process that separates finely ground minerals from one another by causing some to float in a froth and others to sink), gravity, magnetism, or other methods, usually at the mine site, to prepare it for further stages of processing. The production of large amounts of waste material (often very acidic) and particulate emissions have led to major environmental and health concentrations. Additional processing separates the desired metal from the mineral concentrate.

The ingredients or factors shown in Figure 13.1 illustrates that there are various receptor types around the mine site at different times during the life cycle of the mine. These types are detailed below:

- **Construction worker**—may be exposed for short or extended periods depending on role and responsibilities; levels of exposure differ depending on mine's life-cycle stage when work is performed and the location of work relative to source.
- **Outdoor worker**—experiences potential exposure form dust, radiologicals, and hazardous materials.
- **Indoor worker**—experiences either less exposure if in office spaces or potentially more exposure if inside process areas.
- **Off-site tribal practitioner**—assumed that tribal peoples may use traditional hunting and fishing areas for some level of subsistence.
- **Recreational user**—may use lakes, streams, or trails near the mine site or recycling facility and may also boat, swim/wade, bike, hike, camp, hunt, fish or subsist temporarily in the area.
- **Agricultural worker**—may experience more exposure from dusts, noise, or impacted water supply.
- **Trespasser**—exposure depends upon mine site life-cycle state and activity while on-site.
- **Off-site resident**—exposure would depend upon the mine site life-cycle stage and distance from potentially multiple source areas; routes could be air, ingestion of dust or native or gardened plant or animal, ingestion of contaminated water, and dermal contact with soil or water.
- **On-site resident**—exposure would occur after mine land is reclaimed and re-developed for residential use. Routes of exposure could be air, ingestion of dust or native or garden plant of native animal, ingestion of contaminated, water, and dermal contact with soil or water depending on residential concentrations remaining in un-reclaimed source areas or in yard soil if mine wastes were mixed with clean soil and used as fill.
- **Ecological receptors**—aquatic and terrestrial.

Note that direct exposure can occur as a result of direct contact with solid phase mines or process wastes. Therefore, general protections at mine sites are typically required and in place, especially those located on federal lands. For example, fencing is generally required for isolating mine site areas where certain leaching chemicals are used to protect and prevent the direct exposure of the public, wildlife (including migratory birds), and livestock. Indirect exposure to humans can occur through the food chain by, for example, or by consuming vegetables grown in contaminated soils.

In its 1989 risk guidance, EPA stipulated that a completed exposure pathway may contain the following elements:

- source and mechanism for release of chemicals
- transport or retention medium
- point of potential human contact (exposure point) with the affected medium
- exposure route (e.g., dermal contact, ingestion, or inhalation) at the exposure point

Note that if any of these elements is missing, then no human health or ecological risk exists. Again, Figure 13.1 shows a simplified depiction of the ten chemical receptor types.

Contaminant Release and Transport

No two mining operations are identical. Moreover, the environmental behavior from the mining and processing activities involving mine/ore materials, REE-containing minerals, and the mineralogy of overburden and waste rock can vary significantly. Some of the potential effects include alteration of wildlife habitats, erosion, sedimentation, generation of windblown particulates, pollutant loading in groundwater and surface water, losses of chemical solutions from the process areas, and surface subsidence. By and large, the specific areas of concern arise from sediment loading, metal contamination, toxic chemical release, and acidification (USEPA, 2012).

Water Pathway

Surface Water Pathway

Along with mining and processing activities, exploration activities can initially impact surface water and groundwater resources at the site. The potential impacts that could occur during this phase of a mine's life are variable and are influenced by location; impacts in densely forested areas will be different than the impacts to sparsely forested and arid regions. Ore bodies located in more remote locations may require an extensive network of access roads to drilling sites that result in the removal of habitat and alteration of terrain by removing soil and rock to create a stable road bed. Where water bodies exist along these constructed unpaved access roadways, or the drilling sites, pollution to streams and other water bodies can potentially become problematic. Additional runoff from clearing or other land alterations can increase normal stream flow during rainfall events, which increases the potential for downstream impacts from flooding. These types of impacts are more pronounced the closer the access roadways are constructed to the water body and the greater the area cleared.

One of the significant mining activities that can impact surface water bodies and groundwater is the drilling fluids that are used and especially if accidentally released to the environment. Suspended and dissolved solids concentrations would potentially overwhelm and devastate a small stream. Note that the drilling fluids are managed at the borehole, either in a constructed mud box or in a pit. After the borehole is completed, the drilling fluids may be contained in drums for disposal, moved to an on-site waste management area (e.g., a landfill), dispersed at a landfill application unit (i.e., landfarming or land-spreading), or stabilized in the mud pit and buried in place. It is common practice to recycle the drill cutting (i.e., road spreading and on-site construction base material), but additional uses are being tried in some sectors, such as the use of drilling mud and cuttings for use as a substrate for wetland revitalization (USEPA, 2012). During any phase of mining operations, water bodies may receive increased sediment loads from erosion of freshly exposed soils that can cause decreases in available dissolved oxygen (DO) content of waters and increase turbidity and thus decrease light penetration for photosynthesis to occur for aquatic plants.

A natural acidification of runoff water (i.e., acid rock drainage or ARD) from the erosion of rocks surfaces containing sulfides can affect surface water bodies. The releasing of metals from mined materials and surfaces is exacerbated from acid mine drainage (AMD) and neutral mine drainage (NMD). Also, naturally occurring ARD is not to be forgotten or overlooked in this dilemma because the AMD, NMD, and ARD can make for a toxic mix. AMD occurs when oxide ore minerals (metalliferous minerals) are altered by weathering, rainwater, and surface water into oxides or sulfides. Note that AMD is usually not a significant issue for REE deposits; on the other hand, also note that it is the rock that surrounds or overlies the ore body that may contain the sulfide materials that could create AMD. REEs often occur in ores rich in carbonate minerals, which can help buffer any effects of AMD that might occur; however, aquatic systems are very sensitive to changes in pH and increases in alkalinity can also be problematic. While AMD can result in metal toxicity problems, divalent metals are generally less toxic at higher pH and in more mineralized waters associated with NMD. AMD and NMD are collectively referred to as mining-influenced water (MIW). Recall that MIW is defined as any water whose chemical composition has been affected by mining or mineral processing and includes ARD, neutral and alkaline waters, mineral processing waters, and residual waters. MIW can contain metals (including REEs), metalloids, and other constituents in concentrations above regulatory standards. Because the surface area of mined materials is greatly increased, the rate of chemical reactions that generate AMD, or increase alkalinity is also greatly increased. MIW can occur from stockpiles, storage piles, and minded or cut faces that can potentially impact local soil, groundwater, and surface water quality. Typically, drainage from these management areas is controlled, but releases also can potentially happen due to overflows during storm events, liner failure, and other breaches of engineered controls.

Water contamination near lands enriched with REEs is a problem worth attention as well as REEs on aquatic plants. Note that aquatic plants will generally only tolerate narrow fluctuations in the acidity or alkalinity of the natural system until reproductive capacity is diminished or mortally occurs. Erosion of exposed rocks can cause an increase (more acidic) or decrease (more basic) in the hydrogen ion concentration (pH) of the aquatic environment. Prior to mining, oxidation is a function of natural weathering, and any acid generation (i.e., ARD), or increases in alkalinity, occurs very slowly. Soils and riparian areas help to mitigate AMD and MIW, but these buffer zones are usually missing from the mining site unless maintained or constructed. Earlier in the book, it was noted that AMD is not a significant issue that is typically associated with REE deposits. And *typically*, this is the case. However, Gonzales et al. (2014) summarized data about REEs in aquatic ecosystems, suggesting that toxicity depends on the route of supply and chemical form; mining causing the REE release to aquatic environments can follow their route of supply and deliver their chemical form. A good example of REE contamination of a plant species is the aquatic plant *Lemna minor* which is reported by Ippolito et al. (2007) in their research. They found that RRE exposure caused an alteration of roots in *L.*

minor plants treated with REE up to millimolar concentrations. REEs were found to increase both reactive oxygen species and antioxidant defenses and can be interpreted as an indicator or toxicity of REEs for L. minor (Ippolito et al., 2007; Razinger et al., 2007).

Note: If you are not familiar with the aquatic plant *Lemna minor* think duckweed, because that what it is.

Not all the news is bad about the contamination of aquatic ecosystems. Potential impacts to surface waters from mining operations can be mitigated. Maintaining buffer zones between stream areas and areas of exploration and mining activity can help to control runoff to streams. Capturing and containerizing drilling fluids also can mitigate impacts to streams, groundwater, and nearby habitat. Excavated soils, rock, and dry cuttings, either from mine area or drilling activities, can also be controlled to prevent the release of sediments and contaminated runoff.

Okay, plants are one form of life that REEs do or might affect via mining processes the question becomes: how about other forms of life. Consider, for example, microbes, or microorganisms and REE. The truth be known very little research has been done in this area. However, we do know that several microbial species have been reported to be resistant to high levels of REEs in their growth environment and natural and laboratory conditions (d'Aquino and Tommasi, 2017).

Groundwater Pathway

In the hydrogeological cycle (aka water cycle) groundwater and surface water interactions are common. The problem is that at mining sites those interactions are often enhanced. Groundwater can migrate to existing mine pit lakes or evaporation ponds or be recharged from surface mine units. Mine pit water is generally removed to evaporation or treated and discharged to surface waters or injected into the aquifer. It also is often used in the milling processes. Water has a way of finding a way to flow where it wants to—seepage water—especially when it is stored in mine pit lakes or other earthen storage ponds. When the mine water is able to leak from storage it is usually contaminated with chemicals that are conveyed along with the water into alluvial and groundwater flow systems—periods of high precipitation make this problem worse. Groundwater inflow to pit lakes, streams, buried trenches, or surface ditches may also result in the transport of chemicals from subsurface environments to surface waters, including the transfer of chemicals in suspended sediments.

Systems such as pump-back wells are sometimes used to extract a portion of mineralized groundwater or pit seepage water that could migrate off-site and impact neighboring or nearby human or ecological receptors. The effluent is then pumped and released to lined evaporation ponds, resulting in an accumulation of potentially contaminated sediment. Chemical precipitated accumulations can also occur in active pump-back systems and evaporation ponds. Losses of dewatering effluent from pipelines carrying water from mine pits, holding ponds, or from pump-back systems can be a potential source of soil and groundwater contamination (Spellman, 2017a; USEPA, 2012).

Case Study 13.1 presents an example of environmental damage to the aquatic environment caused by mine water release.

Case Study 13.1—South Maybe Canyon Mine Site

In 2011, U.S. Forest Services announced that there is one CERCLA site within the Blackfoot River sub-basin where REE-containing mineral deposits were mine. An Administrative Order of Consent for South Maybe Canyon Mine Site was entered into by the U.S. Forest Service and Nu-West Mining, Inc., in June 1998. The primary reason for the order was the release of hazardous substances, including selenium, from the site into groundwater and surface waters above Idaho state water quality standards. It should be noted that rare earth metals were not identified as hazardous substances that had been released from the site. The South Maybe Canyon Mine was developed for the production of phosphate, and REEs were recovered as a byproduct (Long et al., 2010). This mine has been identified as a possible source of REEs for future development.

Air Pathway

Before discussing air contaminants generic to mining in general and for REEs it is important to ponder the following foundational information first.

Air Contaminants

One of the primary categories of industrial health hazards that miners and affected others must deal with is airborne contaminants. Air contaminants are commonly classified as either particulate or gas and vapor contaminants. The most common particulate contaminants include dusts, fumes, mists, aerosols, and fibers.

Dusts are solid particles that are formed or generated from solid organic or inorganic materials by reducing their size through mechanical processes such as crushing, grinding, drilling, abrading, or blasting.

Industrial atmospheric contaminants exist in virtually every workplace. Sometimes they are readily apparent to workers, because of their odor, or because they can actually be seen. Industrial hygienists, however, can't rely on odor or vision to detect or measure airborne contaminants. They must rely on measurements taken by monitoring, sampling, or detection devices.

Fumes are formed when material from a volatilized solid condenses in cool air. In most cases, the solid particles resulting from the condensation react with air to form an oxide.

The term *mist* is applied to a finely divided liquid suspended in the atmosphere. Mists are generated by liquids condensing from a vapor back to a liquid or by breaking up a liquid into a dispersed state such as by splashing, foaming, or atomizing. *Aerosols* are a form of a mist characterized by highly respirable, minute liquid particles.

Fibers are solid particles whose length is several times greater than their diameter.

Gases are formless fluids that expand to occupy the space or enclosure in which they are confined. Examples are welding gases such as acetylene, nitrogen, helium, and argon; and carbon monoxide generated from the operation of internal combustion engines or by its use as a reducing gas in a heat-treating operation. Another example is hydrogen sulfide which is formed wherever there is decomposition of materials containing sulfur under reducing conditions.

Liquids change into vapors and mix with the surrounding atmosphere through evaporation. Vapors are the volatile gaseous form of substances that are normally in a solid or liquid state at room temperature and pressure. They are formed by evaporation from a liquid or solid and can be found where parts cleaning and painting take place and where solvents are used.

Although air contaminant values are useful as a guide for determining conditions that may be hazardous and may demand improved control measures, the mining managers must recognize that the susceptibility of workers varies.

Even though it is essential not to permit exposures to exceed the stated values for substances, note that even careful adherence to the suggested values for any substance will not assure an absolutely harmless exposure. Thus, the air contaminant concentration values should only serve as a tool for indicating harmful exposures, rather than the absolute reference on which to base control measures.

Routes of Entry

For a chemical substance to cause or produce a harmful effect, it must reach the appropriate site in the body (usually via the bloodstream) at a concentration (and for a length of time) sufficient to produce an adverse effect. Toxic injury can occur at the first point of contact between the toxicant and the body, or in later, systemic injuries to various organs deep in the body. Common routes of entry are ingestion, injection, skin absorption, and inhalation. However, entry into the body can occur by more than one route (e.g., inhalation of a substance also absorbed through the skin).

Ingestion of toxic substances is not a common problem in industry—most workers do not deliberately swallow substances they handle in the workplace. However, ingestion does sometimes occur either directly or indirectly. Industrial exposure to harmful substance through ingestion may occur when workers eat lunch, drink coffee, chew tobacco, apply cosmetics, or smoke in a contaminated work area. The substances may exert their toxic effect on the intestinal tract or at specific organ sites.

Injection of toxic substances may occur just about anywhere in the body where a needle can be inserted but is a rare event in the industrial workplace.

Skin absorption or contact is an important route of entry in terms of occupational exposure. While the skin (the largest organ in the human body) may act as a barrier to some harmful agents, other materials may irritate or sensitize the skin and eyes, or travel through the skin into the bloodstream, thereby impacting specific organs.

Inhalation is the most common route of entry for harmful substances in industrial exposures. Nearly all substances that are airborne can be inhaled. Dusts, fumes, mists, gases, vapors, and other airborne substances may enter the body via the lungs and may produce local effects on the lungs or may be transported by the blood to specific organs in the body.

Upon finding a route of entry into the body, chemicals, and other substances may exert their harmful effects on specific organs of the body, such as the lungs, liver, kidneys, central nervous system, and skin. These specific organs are termed *target organs* and will vary with the chemical of concern (see Table 13.2).

TABLE 13.2
Selected Toxic Contaminants and the Target Organs They Endanger

Target Organs	Toxic Contaminants
Blood	Benzene
	Carbon monoxide
	Arsenic
	Aniline
	Toluene
Kidneys	Mercury
	Chloroform
Heart	Aniline
Brain	Lead
	Mercury
	Benzene
	Manganese
	Acetaldehyde
Eyes	Cresol
	Acrolein
	Benzyl chloride
	Butyl alcohol
Skin	Nickel
	Phenol
	Trichloroethylene
Lungs	Asbestos
	Chromium
	Hydrogen sulfide
	Mica
	Nitrogen dioxide
Liver	Chloroform
	Carbon tetrachloride
	Toluene

Source: Data from Spellman (2017b).

The toxic action of a substance can be divided into short-term (acute) and long-term (*chronic*) effects. Short-term adverse (acute) effects are usually related to an accident where exposure symptoms (effects) may occur within a short time period following either a single exposure or multiple exposures to a chemical. Long-term adverse (chronic) effects usually occur slowly after a long period of time, following exposures to small quantities of a substance (as lung disease may follow cigarette smoking). Chronic effects may sometimes occur following short-term exposures to certain substances.

Fugitive Dust in Mining Operations

Fugitive dust (escaped dust) is an environmental air quality term for very small particles suspended in the air. In mining operations, it is primarily mineral dust that is sourced from mining activities such as digging, drilling, and crushing operations. Direct exposure of humans can occur from inhalation of the fine dusts (i.e., particulates) or by ingestion or dermal contact of contaminated dusts. Particulates or fugitive dust from storage piles, conveyor systems, site roads, or other areas can be transported by wind and may be deposited and accumulated in downwind areas, including surface soils and surface water bodies (e.g., ponds and pit lakes), or be inhaled by site workers and nearby residents. Dust can be an irritant, a toxicant, or a carcinogen, depending on the particles' physiochemistry and can be composed of inorganic and organic chemicals. However, the presence of physical barriers, such as vegetation or structural foundations, may dampen or reduce or block the transport of articles as wind-blown dust. Accumulated mine dust or sediments may become secondary sources of chemicals transported to groundwater through leaching and percolation (Spellman, 2017b).

Aerosol and Chemical Vapors in Mining Operations

Mine workers can be exposed to aerosols from numerous processes, including comminution (i.e., the process in which solids materials are reduced in size by crushing, grinding, and other techniques), re-entrainment (i.e., air being exhausted is immediately brought back into the system through the air intake and/or other openings), and combustion sources. Aerosols in mines are dispersed mixtures of dust and/or chemical-containing water vapor generated by cutting, drilling, and blasting of the parent rock in underground works—the aerosol created has a composition similar to the parent rock. In addition, when the ore is comminuted underground to enable efficient transport out of the mine area aerosols can accumulate in these areas. During the ore mineral liberation and separation, steps dusts settle to other areas such as ventilation systems, work areas, roadways, and nearby surrounding areas. Sometimes aeration ponds using electric aerator devices are used to treat wastewater on a mine site. In operation these aerators disturb the surface of the water and create aerosols; the problem can become worse if surfactants (reduces the surface tension of the water) are used and not properly managed. Aerosols usually accumulate along the perimeter areas of ponds and lagoons and contaminate soils, sediments, surface water, and shallow groundwater through deposition and transport (USEPA, 2012).

Radioactivity

Some level of radioactive material is found in association with many REE deposits; lanthanides and yttrium are recovered primarily from ores and minerals that naturally contain uranium and thorium. As a result, the waste rock and sludges from the extraction of rare earths also contain these radionuclides and are considered Technologically-Enhanced, Naturally-Occurring Radioactive Materials (TENORM). The problem or concern is that TENORM wastes contain radionuclides in concentrations that could give rise to unacceptable radiation levels. The EPA has estimated that levels of naturally occurring radioactivity contained in common rare earth (e.g., monazite, xenotime, and bastnasite) deposits in the United States range from 5.7 to 3,224 picoCuries per gram (pCi/g). Radioactive uranium and thorium can concentrate in mining dusts and sediments that must be managed. Radon gas can also be emitted from these sources. Transport of particulates containing uranium and thorium may occur by any of the transport pathways described earlier. Acidic groundwater and surface water and low concentrations of organic material in soils can contribute to the mobility and transport of radioactive materials. A buildup of sediments deposited by runoff and dusts also can concentrate radioactive materials. Note that external exposure to naturally occurring radiation and radon gas is often limited to soil or waste materials that are within several inches of the ground or pile surface; radioactive materials found deeper in the soil column or accumulated sediments are generally shielded by the top layer of soil. Reduction in the intensity of ionizing radiation with distance from a source (aka in nucleonics as geometric attenuation) limits the external radiation from naturally radioactive materials with no interposed shielding materials to within a few meters (i.e., less than 5 m and often less than 1 to 2 m from the source). Mineral scales can be an issue with radioactivity from mining operations when the scales develop in groundwater recovery wells, holding tanks, aeration ponds, and milling process areas. However, the greatest issue and concern with mining naturally occurring radioactive materials is the inhalation of contaminated dusts.

Tailings storage facilities (TSF), when dry, usually represent the main source of radon/thorium and dust emissions to the environment. When the TSF is wet there is another issue when they receive the bulk of processing water, normally recycled, but some of it could be released into the environment via seepage or overflows due to unusually high rainfall. Special containment arrangements (aka secondary containment) are required for the disposal of tailings generated during chemical and/or thermal processing of uranium and thorium-bearing minerals to ensure that environmentally mobile radionuclides are not released into the surrounding environment. Another control measure that is important is to ensure that the disposal of contaminated equipment and materials is properly monitored and managed. In addition, another factor to consider is the possibility of off-site contamination from trucks and equipment moving off-site.

Minerals, after they have been milled and concentrated are sometimes stockpiled at the mine site prior to transport to the processing sites, and these stockpiles may contain radioactive minerals in concentrations sufficient to produce elevated radiation levels and radon. Therefore, the stockpiles need to be protected against unauthorized access and also against the possibility of the material spreading through wind saltation (i.e., pebble transport). To simplify the management and clean-up operations of stockpiled waste materials containing radionuclides, it is advised to use a concrete pad with signposts stating "supervised" and/or "controlled" (USEPA, 2012).

In many instances, appropriate management is required when dealing with the tailings for the separation and downstream process of minerals that may contain radionuclides in concentrations that could give rise to unacceptable levels of radiation and radon. The disposal of the waste will depend on the method used to process the mineral and on respected levels of radiation and the concentration of radon gas emitted. When separating heavy mineral sands, the possibility of groundwater contamination increases if chemical and/or thermal treatment of the mineral occurs. Radium could be present in the tailings water, which would require removal before being disposed.

Whenever grinding, chemical, and/or thermal treatment of minerals containing radionuclides takes place, additional safeguards must be implemented due to the fact that material equilibrium in both uranium and thorium decay chains may be disrupted. This could result in an increased environmental mobility of radionuclides, such as radium and radon. The cleaning of heavy mineral sand grains and other certain minerals prior to process may produce finely powdered waste (slimes). Because slime wastes may have significant uranium and thorium content, and the disposal as radioactive waste any be required.

Either throwaway or decontamination is required whenever equipment used in downstream processing becomes contaminated by TENORM. Also, attention must be paid toward scale and sludge build-up on the inside surfaces of pipes and vessels used in chemical and thermal processing; these materials might have high levels of radionuclides. Radioactivity can't be ignored and must be mitigated. Case Study 13.2 provides a good example of environmental damage from radioactivity.

Case Study 13.2—Radionuclide in Malaysia

According to a Physorg.com news article (Zappei, 2011), international experts are currently investigating whether a refinery being built by an Australian mining company, Lynas Corp. Ltd., to process rare earth minerals in Malaysia presents any threat of radioactive pollution. Malaysia's last rare earth refinery in northern Perak state was closed in 1992 following protests and claims that it was the source of radionuclides that were identified as the cause of birth defects and leukemia among nearby residents. The refinery is one of Asia's largest radioactive waste cleanup sites. The Pahang plant is meant to refine slightly radioactive ore from the Mount Weld mine in Western Australia, which will be trucked to Fremantle and transported to Malaysia by container ship. Lynas planned to begin operations at the refinery late in 2011 and it is expected to meet nearly a third of world-wide demand for rare earths, excluding China (USEPA, 2012).

Hard Rock Mine Environmental Risks

In this section, generalized risk information is presented based on research and past studies of hard rock mines. Past information gathered by EPA about operations at hard rock mines is applicable to the mining of REEs. This is the case because the geologic environments where other types of metal mines are developed and operated are often similar to geologic settings in which REEs are mined. The point is whether mining for gold, silver, copper, or other metals the mining of REEs has the same environmental risks.

Collected human and ecological risk data (USEPA, 2011), the most current and comprehensive report, was summarized from a subset of 25 National Priority List hard rock mining and mineral processing sites by EPA and is presented here in relation to risks that could also be typical from a hard rock mining operation for REEs. The summary of the contaminant sources, primary transport media or pathway, routes, and receptors is provided in Figures 13.2 and 13.3.

CONTAMINANT RELEASE ➡ CONTAMINANT TRANSPORT ➡ CONTAMINANT FATE

NPL Sites

SOURCES		PATHWAYS	ROUTE	RECEPTORS
Acid mine/rock drainage	Potliners	Groundwater	Dermal contact	Occupational (agriculture)
	Pressing ponds	Sediment	External radiation	Occupational (construction) current
Asbestos fibers	Process fluids	Soil	Ingestion of	Occupational (construction) future
	Process stacks emissions	Surface water	Food-plants	Occupational (dredging) future
Demolition dumps	Quatizite dust slurry	Vadose zone	Food-birds	Occupational (indsurial) current
Deposition	Radioactive waste piles		Food-mammals	Occupational (industrial) future
Dredged sediments	Run-off		Food-terrestiral	Occupational (non-specified) current
Flue dust	Sewage sludge		Food-benthic	Occupational (non-specified) future
Fugitive dust			Food-fish	Recreational-current
Housekeping debris			Food-aquatic plants	Reccreational-hiker
Incinerator ash	Storage tanks		Food-aquatic	Residential-current
Iron-rich Liquid Acid	Sulfate residuals		Inhalation	Residential-future
Metal ore	Waste tailings		Combined rotues	Site visitor
Non-contacting cool	Transformers			Trespasser-current
Water effluent	Treater dust stock piles			Trespaser-future
Ore slimes	Underflow solids piles			
Ore/nodule stockpiles	Unline pits			
Overburden	Waste drums			
Overburden	Waste piles			
	Waste rock			

REMOVAL SITES

SOURCES		PATHWAYS	ROUTE	RECEPTORS
Adits drainage	Overburden	Air	Dermal contact	Occupaitonal
Adits (aka mine tunnel)	Raffinate leachate	Groundwater	Inhalation	Recreational-current
	Tailings	Soil	Ingestion	Residential-current
	Vat leachgagte tailings	Surface water		Trespasser-current
Condenser waste	Waste drums			
Manual/aerial disposition	Waste rock piles			
Mill foundation	Water drainage			
Mill tailings				
Mine waste				

FIGURE 13.2

CONTAMINANT RELEASE ➡ CONTAMINANT TRANSPORT ➡ CONTAMINANT FATE

SOURCES		PATHWAYS	ROUTE	RECEPTORS	
Acid drainage	Ore stockpiles	Food-aquatic invertebrates	Dermal contact	**Birds**	
Adits	Overburden	Food-aquatic plants	Ingestion	American Dipper	Migratory Birds
Airborne emissions	Phossy water	Food-benthic invertebrates	Inhalation	American Kestrel	Mountian Chickadee
ARD	Potliners	Food-birds	Combined routes	American Robbin	Northern Harrier
Process residues		Food-fish		Barn Owl	Omnivorous Birds
Process stacks air emissions		Food-mammals		Belted Kingfisher	Pine Grosbeak
Quartizite dust slurry		Food-plants		Bobwhite Quail	Red-tailed Hawk
Run-off		Food-terrestrial invertebrates		Carnivorous Birds	Red-tailed Hawk
Condenser Waste		Food-terrestrial plants		Cliff Swarrow	Sage Grouse
		Groundwater		Great blue heron	Song Sparrow
		Plants		Horned Lark	Spotted Sandpiper
		Sediment		King Fisher	Waterfowl
Debris	Sludge	Soil		Mallard	Woodcock
Exposed mineralized	Spent mineral waste	Soil invertebrates		**Mammals**	
Bedrock				Carnivorous mammals	Mink
Fuel/oil	Spent ore	Subsurface soil		Coyote	Montane Vole
Fugitive dust	Tailings	Surface soil		Deer	Omnivorous mammals
Leachate	Treater dust stack piles	Surface water		Deer Mouse	Piscivorous Mammals
		Water		Deer Mouse	Rabbits
Manual/aerial desposition	Underflow sands piles			Field Mice	Raccoon
Mine waste	Vat leachate tailings			Herbivorous mammals	Red Fox
Municipal waste	Waste drums			Long-tailed Weasel	Small mammals
Nodule stockpiles	Waste rock			Marked Shrew	Soil invertebrate-feeders
Non-Contacting cool	Wastewater			Meadow Vole	White-tailed Deer
Water effluent	Water drainage			**Other**	
				Amphibians	Plants
				Benthic-macroinvertebrates	Predatory Fish
				Benthic-invertebrates	Rainbow Trout
				Benthic organisms	Sagebrush
				Deepwater habitats	Soil invertebrates
				Fish/pets	Terrestrial invertebrates
				Future-aquatic organisms	Terrestrial organisms
				Future-aquatic invertebrates	Terrestrial plant community
				Future-wildlife	Terrestrial soil
				Herbivores	Thickspike Wheatgrass
				Macroinvertebrates	Transient organisms
				Periphyton community	Vegetation

FIGURE 13.3

The sample set of NPL sites represents those listed on the NPL post-1980, and it is thought that these sites likely represent the conditions that could be found at modern mine sites. In simple extrapolation, it can be shown that similar source, fate, and transport scenarios could (and can be) be related to REE mining and processing facilities.

From a previous study by USEPA in 1995 of the risks to human health and the environment from hard rock mines, it was concluded that 66 hard rock mining cases illustrate that significant human health and environmental damages were caused by the management of wastes from mining and mineral processing, particularly placement in land-based units. Molycorp Mineral rare earth mine in Mountain Pass, CA, was one of the sites included in the study. It was the wastes subject to the Bevill Amendment that was found to be the cause of damages. These damages occurred in all hard rock mining sectors and across all geographic regions of the United States. Excerpts from the USEPA study are summarized by type of impacts in Table 13.3.

In a 2004 USEPA review of 156 hard rook mine sites, the results showed that approximately 30% (or about 45) of the sites had problems with AMD. The report also speculated that AMD occurred most often in EPA Regions 8, 9, and 10. EPA took up the National Wildlife Federation's mantra that sites where AMD is present should be carefully monitored as "the presence of acid mine drainage is either underestimated or ignored until it becomes evident, at which time the costs often exceed the operator's financial resources, leading to bankruptcy of abandonment of the site in many cases" (USEPA, 2004). Again, as previously stated and given the typical geochemistry of these deposits, it is not anticipated that the mining of REE will typically produce AMD.

REE-RECYCLING PATHWAYS AND EXPOSURES

Exposure to hazardous materials is probably minimal during the collection of items to be recycled and if exposure actually does occur it will likely result from either inhalation or dermal exposures to materials released from damaged items. In the same way, and when done properly, manual dismantling is likely to have a low potential for worker risks resulting from exposure to hazardous materials. Mechanical

TABLE 13.3
Frequency of Various Types of impacts form CERLA Sites (USEPA, 1995)

Type of Impacts	Portion of Damage Cases (Total NPL Sites = 66)
Surface water contamination	75% of cases
Groundwater contamination	65%
Soil contamination	50%
Human health impacts	35%
Flora and fauna damage	25%
Air dispersion or fugitive emissions	20%

dismantling and shredding can generate dust containing hazardous components. If not properly controlled, the dust can result in inhalation or dermal exposures to workers. Dust control is particularly important when the items being shredded contain brominated flame retardants because high temperatures during shredding could result in the formation of dioxins (Schluep et al., 2009).

Whenever nitric acid or aqua regia is used in leaching processes nitrogen oxide or chlorine gases can be released and therefore must be controlled to prevent human and environmental impacts. In other processes that use strong acids or bases, safe handling of chemicals and disposal of resulting waste streams is important to protect workers and the environment. Thermal processes used for recycling can result in air emissions, liquid wastes, and solid waste streams. As for mineral processing, TENORM is a safety and environmental concern related to the recycling of metals. Once again, proper controls and handling are necessary to prevent human exposures and environmental impacts.

As reported by Schuler et al. (2011), when compared with primary processing, recycling of REEs will provide significant environmental benefits with respect to air emissions, groundwater protection, acidification, eutrophication, and climate protection. Moreover, this report states that the recycling of REEs will not involve the majority of the impacts from mining operations and the impacts that can result from radioactive impurities, as is the case with the primary production of metals. As previously noted a large percentage of REE-containing materials are shipped to developing countries where they are recycled using informal processes. These informal operations can include manual dismantling, open burning to recover metals, de-soldering of printed wiring boards over coal fires, and leaching in open vessels. In almost all cases, these operations are uncontrolled and can lead to human exposure and extensive environmental damage. Moreover, the processes used are generally less efficient and result in lower materials recovery than from more formal or established methods. These uncontrolled REE recycling and metal recovery operations could be prevented if environmental regulations were in place in the United States.

HUMAN HEALTH AND ECOLOGICAL EFFECTS FROM EXPOSURE TO REE

At the present time (2022), only limited toxicological or epidemiological data are available to assess the potential human health effects of REEs. Research is needed to better understand sources, environmental behavior, ecotoxicology, and human epidemiology (Gwenzi et al., 2018). From the limited toxicological and epidemiological data obtained to date, we have some understanding of the impact or exposure to REEs, however, primarily in their mixture form, rather than individual elements. From this limited information we have identified respiratory, neurological, genotoxicity, and mechanism of action studies have been identified. In this section, what we now know is discussed but keep in mind that much of the data was compiled by Chinese investigators and a lot of detailed information is not available in English, other than that in a few abstracts.

So, before moving on let's do a bit of review to set the stage for what follows. Recall that REEs are broadly grouped into "light" (La, Ce, Pr, Nd, Sm, Eu, and Gd) and "heavy" (Y, Tb, Dy, Ho, Er, Tm, Yb, and Lu) classes (Wells and Wells, 2001; USEPA, 2012). For any given lanthanide, soluble forms include chlorides, nitrates, and sulfates, while insoluble forms include carbonates, phosphates, and hydroxides. The larger, lighter (smaller atomic number), and less soluble ions have been observed to primarily deposit in the liver, while the smaller, heavier (larger atomic number), and more soluble ions are similar in ionic radius to divalent calcium and distribute primarily to bone (Wells and Wells, 2001; USEPA, 2012). Because distinguishing individual lanthanides is analytically challenging, it has been difficult to discern the effects of the individual lanthanides—both in human cases and animal studies. Recall that "lanthanides" refers to 15 elements with atomic numbers 57 through 71: lanthanum, cerium, praseodymium, neodymium, promethium, samarium, europium, gadolinium, terbium, dysprosium, holmium, erbium, thulium, ytterbium, and lutetium. The term "rare earths" refers to the lanthanide series plus yttrium (atomic number 39) and scandium (atomic number 21) (Kirk-Othmer, 1995). Okay, now that we have cleared the air concerning lanthanum it is important to point out that in addition, the co-occurrence of radioactive lanthanides, thorium isotopes, and silica dust has complicated the interpretation of toxicity—especially with regard to human exposure (Palmer et al., 1987; USEPA, 2012).

One topic of debate has been the inhalation of REEs and their pulmonary toxicity affecting human health. This is especially the case with regard to the relative contributions of radioactive contaminants versus stable elements in the development of progressive pulmonary interstitial fibrosis (Haley, 1991; USEPA, 2012). Specifically, although it is understood that stable REE compounds can produce a static, foreign-body-type lesion consistent with benign pneumoconiosis it is uncertain whether they can also induce interstitial fibrosis that progresses after the termination of exposure. Human inhalation toxicity data on stable REEs mainly consist of case reports on workers exposed to multiple lanthanides (USEPA, 2012).

Using its Integrated Risk Information System (IRIS) and Provisional Peer-Reviewed Toxicity Values (PPRTV) programs USEPA reviewed the human health toxicity for a few REEs. Human health benchmark values (where derived) and background toxicity information are presented alphabetically by REE below. The toxicity of dysprosium, erbium, europium, holmium, lanthanum, scandium, terbium, thulium, ytterbium, and yttrium have not been reviewed by EPA or pertinent data is not yet available.

- **Cerium**—a 2009 IRIS study and assessment on cerium oxide and cerium compounds using available human and animal studies demonstrated that ingested cerium may have an effect on cardiac tissue (endomyocardial fibrosis—fibrotic thickening of the endocardium) and hemoglobin oxygen affinity; however, data were insufficient to derive an oral Reference Dose (RfD). An inhalation Reference Concentration (RfC) of 9×10^{-4} mg/m^3 was derived based on the increased incidence of alveolar epithelial hyperplasia in the lungs of rats. The lung and lymphoreticular system effects reported in the principal study are consistent with effects observed in humans, which were characterized by the accumulation of cerium particles in the lungs and lymphoreticular system and histologic effects throughout the lung. Data are unavailable regarding the carcinogenicity of cerium compounds in humans or experimental animals (USEPA, 2012).

Additional primary cerium toxicity studies were conducted by various researchers and their objectives and key findings of a few of the important ones are briefly summarized here.

Gomez-Aracena et al. (2006) examined the association between toenail cerium levels and risk of first acute myocardial infraction (AMI) in a case-control study in Europe and Israel. Key findings: cases had significantly higher levels of cerium than controls after various adjustments. Results suggest that cerium may possibly be associated with an increased risk of AMI. Moreover, cerium was positively associated with low socio-economic status, smoking, mercury, zinc, and scandium.

McDonald et al. (1995) described a male patient's respiratory effects; the patient had a chronic history of optical lens grind, an occupation associated with exposure to cerium oxide. Key findings: patient present with progressive dyspnea and an interstitial pattern on chest X-ray; open lung biopsy showed interstitial fibrosis, while scanning electron microscopy demonstrated number particulate deposits in the lung (most containing cerium). This is one of the first cases to describe ER pneumoconiosis associated with pulmonary fibrosis in the occupational setting of optical lens manufacture.

Palmer et al. (1987) evaluated cerium in a vitro cytotoxicity assay system using rate pulmonary alveolar macrophages. Both the soluble chloride form and their insoluble metal oxides were studied. The results suggest that rare earth metal fumes should be considered as cytotoxic to lung tissue and potentially fibrogenic.

- **Gadolinium**—in a 2007 PPRTV document, minimal effects on body weight gain and liver histology were reported following ingestion of gadolinium in rats. Pulmonary histopathological changes with manifestation that included decreased lung compliance and pneumonia leading to mortality were observed in mice and guinea pigs subchronically exposed to gadolinium oxide via inhalation. Data were insufficient to derive any quantitative health benchmarks. Gadolinium was assigned a weight-of-evidence description of "inadequate information to assess carcinogenic potential" (USEPA, 2012).

Additional primary cerium toxicity studies were conducted by various researchers and their objectives and key findings of a few of the important ones are briefly summarized here.

Bussi and Morisetti (2007) evaluated an in vivo gadolinium release for three magnetic resonance imaging (MRI) contrast agents estimating gadolinium content in liver, kidneys, spleen, femur, and brain after single or repeated intravenous administration to rats. Gadolinium acetate (GdAc) was used as a positive control. Key findings: no blood chemistry, hematology, or histopathology changes were seen with the tested contrast media, whereas increased white blood cell count and serum cholesterol were found after GdAc.

Perazella (2009) examined the possibility of nephrotoxicity from gadolinium-based contrast (GBC) in humans. Key findings: reports of a rare systemic fibrosing condition called nephrogenic systemic fibrosis (NSF) were linked to exposure of patients with advanced kidney disease to certain GBC agents. Only patients with advanced acute or chronic kidney disease were found to be at risk for developing NSF.

Sharma (2010) assessed the contrast agent gadolinium toxicity on mice skin by measuring regional epidermal thickening and hair follicle width resulting from delayed gadolinium contrast MRI. Key findings: gadolinium treatment showed skin toxicity as epidermis thickening due to the use of high concentrations of gadolinium in microimaging.

Yongxing et al. (2000) examined the genotoxicity of trivalent gadolinium in human peripheral blood. Key findings: micronuclei frequency increased in a dose-dependent manner upon exposure to the rare-earth element. Significant differences in single-stranded DNA breaks and unscheduled DNA synthesis.

Haley et al. (1961) examined the effects of $GdCl_3$ in food of rats for 12 weeks. Body weight, hematology, and histology were assessed. Key findings: minimally decreased body weight gain and liver histological alterations were reported in male, but not female, Gd-treated rats.

- **Lutetium**—USEPA (2012) reported that in a 2007 PPRTV document a subchronic oral provisional RfD (p-RfD) of 9×10^{-4} mg/kg/day was derived based on a stand-alone no observed adverse effect level (NOAEL) in mice; there are no data to indicate the toxicological endpoint(s) target organ(s) (i.e., nervous system skin, lungs, liver, kidneys, reproduction, and so forth) or oral exposure to lutetium. A study by Haley (1991) provided a comprehensive assessment of human and animal data. The data concluded that the evidence suggested that inhalation exposure to high concentrations of stable REEs can produce lesions comparable with pneumoconiosis and progressive pulmonary fibrosis and that the potential for inducing these is related to chemical type, physiochemical form, and dose and duration of exposure. Note that data were insufficient to derive an inhalation RfC. Luteum was assigned a weight-of-evidence description of "inadequate information to assess carcinogenic potential" (USEPA, 2012).

  Haley et al. (1964) examined the effects of $LuCl_2$ in food of rats for 90 days. Body weight, hematology, and histology were assessed. No exposure-related histopathological or other changes were observed.
- **Neodymium**—based on a free-standing NOAEL in rats (no effects on body weight, hematology, and histopathology) a subchronic oral p-RfD of 5×10^{-1} mg/kg/day was derived for neodymium. Data were insufficient to derive an inhalation RfC. Neodymium was assigned a weight-of-evidence description of "inadequate information to assess carcinogenic potential" (USEPA, 2012).
- **Praseodymium**—in a 2009 PPRTV document, a subchronic oral p-RfD of 5×10^{-1} mg/kg/day was derived for praseodymium based on a freestanding NOAEL in rates (no effects on body weight, hematology, and histopathology) (subchronic p-RfD = 8×10^{-1} mg $PrCl_3$/kg/day). Data were insufficient to derive an inhalation RfC. Praseodymium was assigned a

weight-of-evidence description of "inadequate information to assess carcinogenic potential" (USEPA, 2012).

- **Promethium**—although a 2007 PPRTV document (cite by an USEPA (2012) report) was prepared for promethium, no human health benchmarks were derived due to lack of data. Promethium was assigned a weight-of-evidence description of inadequate information to assess carcinogenic potential.
- **Samarium**—in a 2009 PRTV document, the lowest observed adverse effect level (LOAEL) was reported for increased relative pancreas and lung weights and increased malondialdehyde concentrations in liver tissues of rats exposed to samarium nitrate in drinking water. Data suggest that different chemical forms of samarium have different toxic potencies. Data were insufficient to derive an inhalation RfC. Samarium was assigned a weight-of-evidence description of "inadequate information to assess carcinogenic potential" (USEPA, 2012).

THE BOTTOM LINE

Although much of the information provided in this book is dated (mostly circa 2012), it is what is currently available. Presently, because of the increasing value and usage of REE in modern technologies, research efforts are on-going in not only studying ways in which REE can be further applied to emerging technologies but also studies are in progress to determine their potential human health and environmental impacts. The good news is that mining, extraction, use, and reuse of REEs is dynamic; meaning, that we are at the threshold of learning what it is that we do not know about what we do not know about REE in all pertinent respects.

REFERENCES

Bussi, S.X.F. and Morisetti, A. (2007). Toxicological assessment of gadolinium release from contrast media. *Experimental and Toxicologic Pathology* 58: 323–300.

d'Aquino, L. and Tommasi, F. (2017) Rare earth elements and microorganisms. In G. Pagano (ed.) *Rare Earth Elements in Human and Environmental Health*. Singapore: Pan Stafford Publishing, pp. 111–128.

Gomez-Aracena, J., Riemersma, R.A., Gutierez-Bedmar, M., Bode, P., Kark, J. D., Garcia-Rodriguez, A., Gorgojo, L., Van't Veer, P., Fernandez-Crehuet, J., Kok, F.J. and Martin-Moreno, J.M. (2006). Toenail cerium levels and risk of a first acute myocardial infarction the EUROMIC and heavy metals study. *Chemosphere* 64: 112–120.

Gonzales, D., et al. (2014). Suppression of medicator is regulatory by cdk-8. *Proceedings of the National Academy of Sciences of the United States of America* 111(7): 2500-2505.

Gwenzi, W., Mangan, L., Danha, C., Chaukura, C., Dunjana, N. and Sanganyado, E. (2018). Sources, behaviour, and environmental and human health risks of high-technology rare earth elements as emerging contaminants. Accessed 01/18/2022 @ https://pubmed.ncbi.nlm.nih.gov/29709849.

Haley, P.J. (1991). Pulmonary toxicity of stable and radioactive lanthanides. *Health Physics* 61: 809–821.

Haley, T.J., Komesu, M., Efros, et al. (1964). Pharmacology and toxicology of lutetium chloride. *Journal of Pharmaceutical Sciences* 53: 1186–1188.

Haley, T.J., Raymond, K., Komesu, N., et al. (1961). Toxicological and pharmacological effects of gadolinium and samarium chlorides. *British Journal of Pharmacology* 17: 526–532.

Ippolito, M.P., Paciolla, C., d'Aquino, L., Morgana, M. and Tommasi, E. (2007). Effect of rare earth elements on growth and antioxidant metabolism in Lemna minor I., *Caryologia* 60: 125–128.

Kirk-Othmer. (1995). Lanthanides. In *Kirk-Othmer Encyclopedia of Chemical Technology*, 4th ed. Volume 14: Imaging Technology To Lanthanides. New York: John Wiley & Sons, pp. 1091–1115.

Long, K.R., Bradley, S., et al. (2010). The principal rare earth elements deposits of the United States—a summary of domestic deposits and a global perspective. USGS Scientific Investigation Report 2010–5220, Reston, Virginia. Accessed 01/20/2022 @ http://pubs.usgs.gov/sir/2010/5220/.

McDonald, J.W., Ghio, A.S.J., Sheenan, C.E., Bernhardt, P.F. and Roggl, V.L. (1995). Rare earth (cerium oxide) pneumoconiosis: analytical scanning electron microscopy and literature review. *Modern Pathology* 8(8): 859–865.

Palmer, R.J., Butenhoff, J.L. and Stevens, J.B. (1987). Cytotoxicity of the rare earth metals cerium, lanthanum, and neodymium in vitro: comparisons with cadmium in a pulmonary macrophage primary culture system. *Environmental Research* 43(1): 142–156.

Perazella, M.A. (2009). Current status of gadolinium toxicity in patients with kidney disease. *Clinical Journal of the American Society of Nephrology* 4(2): 461–469.

Razinger, M. et al. (2007). REEs *L. minor* real-time in vivo visualization of oxidation stress. *Central European Journal of Biology* 2(3): 351–363.

Schluep, M., Hagelueken, C., et al. (2009). Recycling from e-waste to resources. United Nations Environment Programme & United Nations University, July. Accessed 01/11/2022 @ http://www.unep.org/PDF/PressReleases/E-Waste_publication_screen_final-verison-sml.pdf.

Schuler, D., Buchert, M., et al. (2011). Study on rare earths and their recycling. OKO-insititut e v., January.

Sharma, R. (2010). Gadolinium toxicity: epidermis thickness measurement by magnetic resonance imaging at 500 MHZ. *Skin Research and Technology* 16(3): 339–353.

Spellman, F.R. (2013). *The Handbook of Environmental Health.* Lanham, MD: CRC Press.

Spellman, F.R. (2017a). *The Science of Environmental Pollution*, 3rd ed. Boca Raton, FL: CRC Press.

Spellman, F.R. (2017b). *Industrial Hygiene Simplified.* Lanham, MD: Governmental Institutes Press.

USEPA (1995). Human health environmental damages form mining and mineral processing waste: technical background document supporting the supplemental proposed rule apply phase iv land disposal restrictions to newly identified mineral processing wastes. Washington, DC: EPA Office of Solid Waste.

USEPA (2004). Evaluation report: nationwide identification of hardrock mining sites. Office of Inspector General, EPA Report Number 2004-P-00005. Accessed 01.20/2022 @ http://wwwepa.gov/org/reprots/2004/20040331-2004-p-00005.Pdf.

USEPA (2009). Conceptual site model for the Yerington Mine Site. Lyon County, Nevada. EPA Office of Superfund Programs—Region. Accessed 01/20/2022 @ http://yosemite.epa.gov/r9/sfund/r9sfddocw, nsf/cf0fac722e32d408882574260073faed.

USEPA (2011). REEs National Emissions Inventory Data. Accessed 12/12/21 @ https://www.epa.gov/air-emissions-inventories/2011

USEPA (2012). Rare earth elements: a review of production, processing, recycling, and associated environmental issues. EPA 600/R-12/572. Accessed 12/23/2021 @ www.epa.gov/ord.

U.S. Forest Service (2011). South Maybe Canyon mine project: request for comment on report to improve water quality at maybe canyon mine. United State Department of Agriculture. Accessed 01/20.2022 @ http://www.fs.usda.gov/wps/portal/fsinternet/lut/p/c4/04.

Wells, W.H. and Wells, V.L. (2001). The lanthanides, rare earth metals. In E. Bingham, B. Cohrssen and C.H. Powell (eds.) *Patty's Industrial Hygiene and Toxicology*, 5th ed., volume 3. New York: John Wiley and Sons, Inc., pp. 423–458.

Yongxing, W., Xaogrong, W. and Zichen, H. (2000). Genotoxicity of lanthanum (III) and gadolinium (III) in human peripheral blood lymphocytes. *Bulletin of Environmental Contamination and Toxicology* 64: 611.

Zappei, J. (2011). Malaysia reviews safety of rare earth plant. Physorg.com On-line E-zine. Accessed 01/20/2022 @ http://www.Physorg.com/news/2011-04-malaysia-safety-rare-earth.html.

Index

Note: **Bold** page numbers refer to tables and *italic* page numbers refer to figures.